Dietmar Geyer Collecting Writing Instruments

Dietmar Geyer

Collecting Writing Instruments

- From the Flint Tool to the Stylus -
From the Quill Pen to the Fountain Pen and Felt-tip Marker

Schiffer Publishing Ltd

1469 Morstein Road, West Chester, Pennsylvania 19380

Thanks for advice and help

In the writing of this book, the following persons have contributed information, and lent materials and photographic documentation:

Pierre Haury, Paris
Stephen F. Hull, Marlow
Bruno Künzler, Hannover
Philip Poole, London
Peteris Seja, Alfter
Mel Strohminger, Baltimore
Stefan Wallraffen, Cologne

as well as the following institutions and corporations and their staff members:

A. T. Cross, Mainz, Hermann Kirch
German Museum, Rolf Guthsmann
edding Aktiengesellschaft, Ahrensburg, Angelika Ledermann
A. W. Faber-Castell, Stein, Peter Schaffhauser
Geha-Werke GmbH, Hannover, Burghard Kiebe
Wholesale Organization of German Office Supply Companies, Bergisch Gladbach, Wolfgang Römer
C. Josef Lamy GmbH, Heidelberg
Merz & Krell, Grossbieberau, Patrick Politze
Montblanc, Hamburg, H. G. Polenz
Parker Pen GmbH, Baden-Baden, Katja Nagel-Ilkhanipur
Pelikan AG, Hannover, Werner Spieker
Registra AG (Waterman), Egg, Josef Thüring
rotring-werke Riepe KG, Hamburg, Rosemarie Richter
Schwan-Stabilo Schwanhäusser GmbH & Co., Nürnberg, G. C. Flechtner
Staedtler Mars GmbH & Co., Nürnberg, Dr. Geiger

Once again, hearty thanks to you all.

Translated by Dr. Edward Force,
Central Connecticut State University

Edited by Douglas Congdon-Martin.

Library of Congress Catalog Number: 90-61508.

Printed in the United States of America.
ISBN: 0-88740-277-1

This book originally published under the title,
Schreibgeräte Sammeln,
by Verlag Laterna Magica, D-8000 München 71,
© 1989. ISBN: 3-87467-391-X.

Published by Schiffer Publishing, Ltd.
1469 Morstein Road
West Chester, Pennsylvania 19380
Please write for a free catalog.
This book may be purchased from the publisher.
Please include $2.00 postage.
Try your bookstore first.

Contents

From the flint tool to the reed pen — the history of writing 6
Calligraphy—the art of beautiful writing 14
The Ink 18
Ink Production Around the Turn of the Century 21
Pelikan 22
The Quill Pen 28
Producing Good Writing and Drawing Pens 30

The Steel Pen 32
Picture Postcards 40
Handwriting 41
Soennecken 42

The Pencil 46
The Pencil in Literature 50
How a Pencil Is Made 52
Faber-Castell 54
J. S. Staedtler 58
Schwan-Stabilo 62

The Fountain Pen: The Early Years 64
The Fountain Pen Pioneer: L. E. Waterman 68
Waterman 72

The Fountain Pen: The Breakthrough 74
George S. Parker—From Teacher to Inventor 74
How It Became a Part of Everyday Life 76
Parker 78
Protected by Patent 82
W. A. Sheaffer—The clever Jeweler 84
Sheaffer 86

The Fountain Pen: The Golden Years 88
Presentation of Writing Instruments 93
Wahl-Eversharp 94
Kaweco 96
Montblanc Simplo 98

From the Stylo to the Indelible Pen 102
rotring 104

The Fountain Pen: Decline and Rise 106
The Mobile Advertisement 112
Lamy 114
Geha 116
The Geha Reserve Tank 116
Historic Fountain Pens 118

Fountain Pen: Types and Materials 126
The Materials 127
- Hard Rubber 127
- Galalite 128
- Bakelite 128
- Celluloid 128
- Lucite 129
- Thermoplastics 129
- Metals 129
- Stainless steel 130
- Engraved metals 130
- Diamond cut 130
- China lacquer 131

Thomas de la Rue (Onoto, Pelikan) 132
Industrial reports: From Raw Metal to Gold Pen 134
The Pen Gives Handwriting Character 136
The Ink Passage—Heart of the Fountain Pen 138
The Filling Systems 139
- The Insertion Pen 139
- The Safety Pen 140
- The Tablet Pen 140
- The Eyedropper Pen 140
- The Tube Pen 141
- The Lever Pen 142
- The Pressure Pen 143
- The Piston Pen 143
- The Membrane System 144
- The Nylon Bag 144
- The Capillary System 145
- The Ink Cartridge 145

Mabie, Todd & Co. (Swan) 146
Decoration 148
- Lavish Decoration (1860 to 1920) 148
- Clear Art Deco Form (1920 to 1929) 148
- Streamline Form (1929 to 1939) 149
- Modern Styles (1939 to 1970) 149
- The Super-Slim Types (since 1970) 149

Caps and Clips 150
Conway Stewart 152
Industrial Report: A Fountain Pen Is Made 154

The Ballpoint Pen 156
The Technology of the Ballpoint Pen 160
A. T. Cross Company 162
How a Ballpoint Pen Is Designed:
The Colani Example 163
Senator 165
Almost a Museum: 6500 Ballpoint Pens under one roof 166

Felt and Fiber Tip Markers 168
Edding 170
How an Ideal Pen Should Be Designed 171

Historic and Modern Pen Brands 172
Early German Brands (no longer in production) 172
Other Brands 172
- Germany 172
- Belgium 173
- France 173
- Italy 173
- Great Britain 173
- Switzerland 173
- Spain 173
- Japan 173
- Pilot 174
- Platinum 174
- Sailor 174
- USA 174

Index 175

From the Flint Tool

"Who writes, remains"
Proverb

Paleolithic drawing of a mammoth.

As if bewitched, a dozen pairs of eyes stared at the wall, only faintly lit by burning branches, of a cave near the present Spanish city of Altamira. Anxiously, the members of the tribe clung to each other, ready to run at any time. For what happened here three thousand years ago had never before been seen by any of them. With skillful movements, the old magician dipped his index finger in a sticky fluid of soil and ocher and made bison, deer and people appear on the cold stone wall. It took a long time until these ice-age people understood: the lines and strokes told a story. Yes, just as shown on the stone wall, they drove the hunted animals on, surrounded them and killed them with long spears. Again and again, the nomadic hunters came back to the "picture cave" and read anew the hunting stories which they hoped would bring them luck.

In such a way, some 10,000 to 60,000 years before our system of counting the years, picture-writing may have had its beginning in the early history of humankind. After thinking and speaking, primitive man succeeded in capturing the spoken word with the help of pictorial symbols. No doubt writing is one of humanity's greatest inventions. The achievements of technology, including computers and space travel, pale in comparison to it. Without writing and the writing materials needed for it, the cultural development of humankind would not have been possible.

The flint tool with which the scratchings of animals were imitated may well have been the first "writing instrument." In the "picture caves" of Lascaux in southern France and Altamira in northern Spain, hundreds of scrapers, gravers and other scratching tools have been found. Primitive paint was applied, at first with the fingers, then with brushes of thin tree stems with feathers stuck into them. For thousands of years, pictures remained the only written language. Gradually becoming more strongly stylized, picture writing developed into symbol writing,

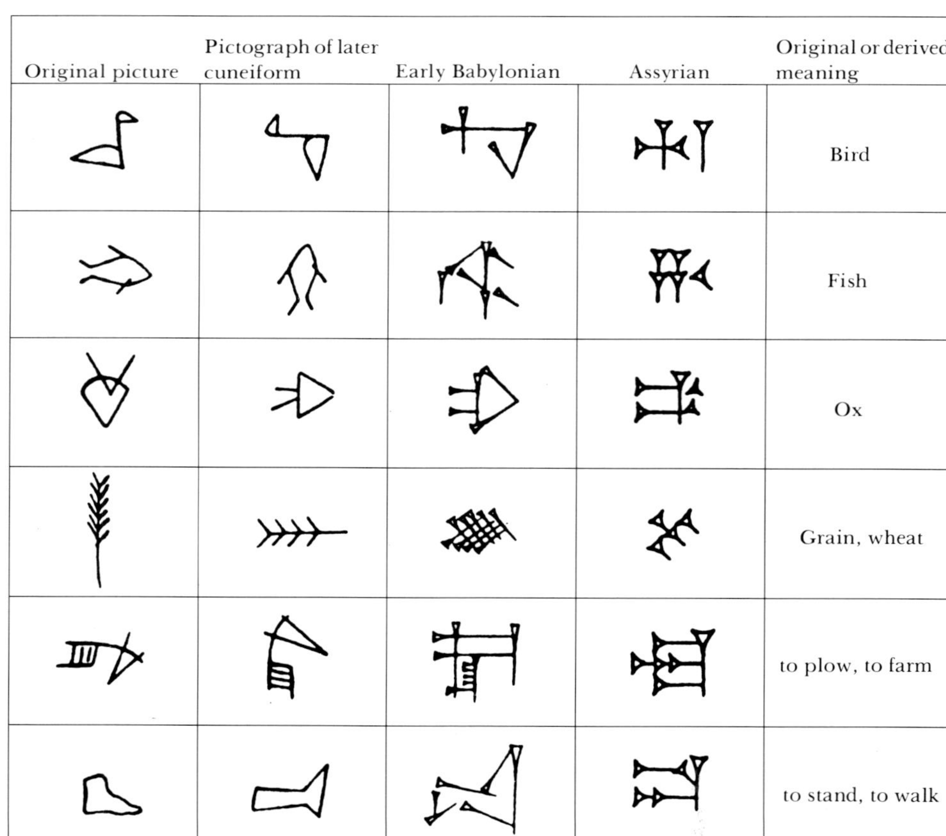

Cuneiform was derived from pictorial symbols.

— to the Reed Pen

pictographs which could be written quickly with as few strokes as possible. A noteworthy example of this was provided by the Sumerians in the land between the Tigris and Euphrates rivers. Some 3500 years ago, they scratched and pressed writing symbols into moist clay tablets with a three-cornered stylus of swamp reeds. From some 2000 original writing symbols cuneiform writing developed, with about 600 symbols of sounds. The Sumerians invented, among other things, not only money but also the practice of lending it for interest. Their relatively highly developed economic system required a correspondingly high cost of government. The air-dried or burned clay tablets proved to be lasting records of transactions, salary records and correspondence. One of the best-known cuneiform tablets is a medical work. In it a collection of prescriptions was recorded 4000 years ago: "Pulverize the seeds of a carpenter's plant, the sticky resin of the markazi plant, and thyme, dissolve in beer, have the man drink it . . ."

At almost the same time, and in a similar way, writing was born in the fruitful Nile valley. The Egyptian writing originated in the scratched symbols with which the farmers marked the height of flood waters on stone walls. Pictorial symbols developed from them that were scratched or chiseled in the stone. The Greeks later named the Egyptian writing "hieroglyphics." which meant "sacred chiseled writing." The oldest writing symbols on the pyramids consisted of vertical lines whose writing ran from left to right. The basic elements of the Egyptian writing were stylized pictures: Waves meant water, thefront of a lion meant the beginning, his rear meant the end. Later ideographs were added to these pictographs as symbols for various concepts, and the system was completed by phonetic symbols—24 symbols for individual consonants.

The pharaohs had symbols engraved in the stones of temples and grave markers, so that posterity would be informed of the power and glory of the god-kings. A handier writing material had to be found for everyday use, however. Instead of the chisel, the graver of bone or metal was used to write on a tablet of wax-covered wood. The scribe, unlike the stonemason, could make the writing disappear by rubbing it with the widened end of the graver. The farther governmental power extended, the greater grew the need for educated official scribes. It was a respected, well-paid profession in ancient Egypt, as one can tell from the advice of a dignitary to his son: "Love writing as your mother . . . then you can protect yourself from hard work of any kind and become a highly respected official . . ."

The famous Rosetta Stone was found by Napoleon's Grand Army in 1799 during their Egyptian collion mde the first step in decipherment by comparing the three written languages: Hieroglyphics, Demotic, and Greek.

Pictures and pictorial symbols chiseled in stone tell posterity of the might of the Pharaohs.

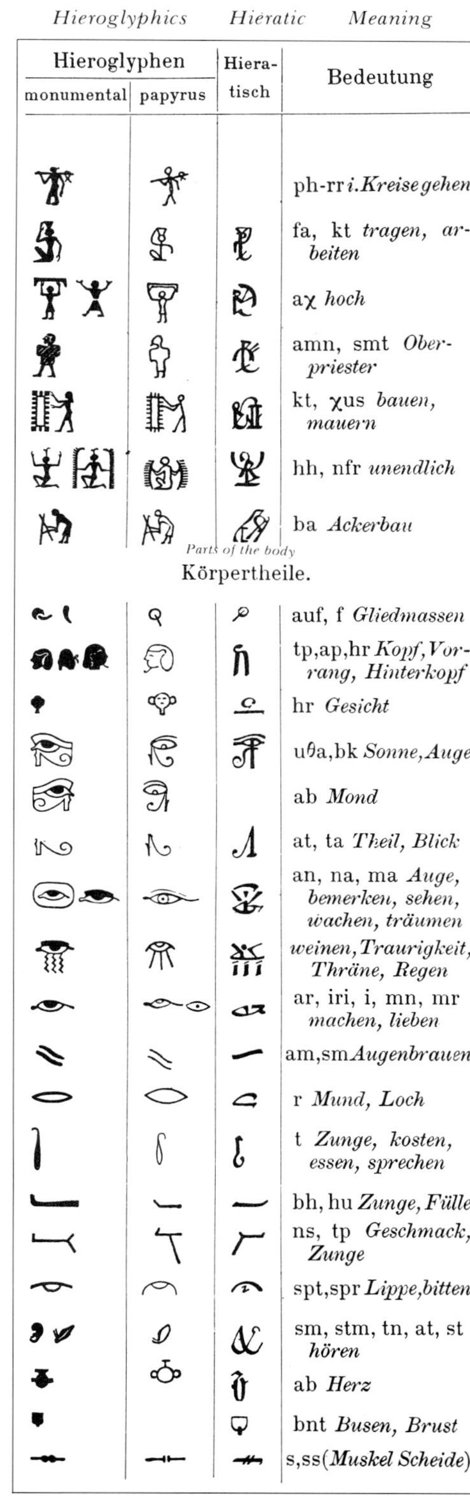

Stylized pictures formed the Egyptian hieroglyphics.

Papyrus

With the growing use of written records came the need for a quickly and flowingly written script. To be sure, the hieratic script of the priests was a simplification, but the essential new development consisted of the inventions of writing and drawing inks and of papyrus. Now it was possible to write things down with the required speed and yet legibly and lastingly.

The raw material for papyrus, from which our word "paper" was derived, grew in front of the Egyptians' doors in the swampy standing waters of the Nile. The pith was removed from the papyrus stalks and the hulls cut into slim lengthwise strips. Placed in layers and pressed together, the remaining plant juices cemented it all together in the form of a roll of writing "paper." The resulting sheets were attached together to lengths of up to forty meters. There were no problems in storing such gigantic sheets. Papyrus was simply rolled up.

Over the millennia, black and red remained the favorite colors for writing ink. Black ink was made by mixing soot, water, and gum. It was required for written

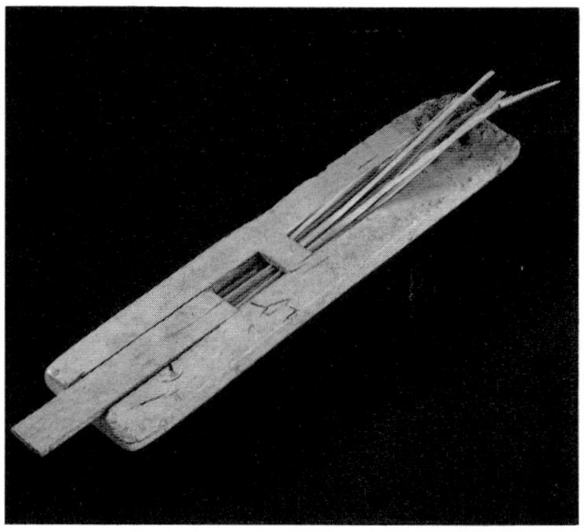

Egyptian writing palette with wells for red and black ink. The writing instruments, five reeds, are held in place by a sliding cover.

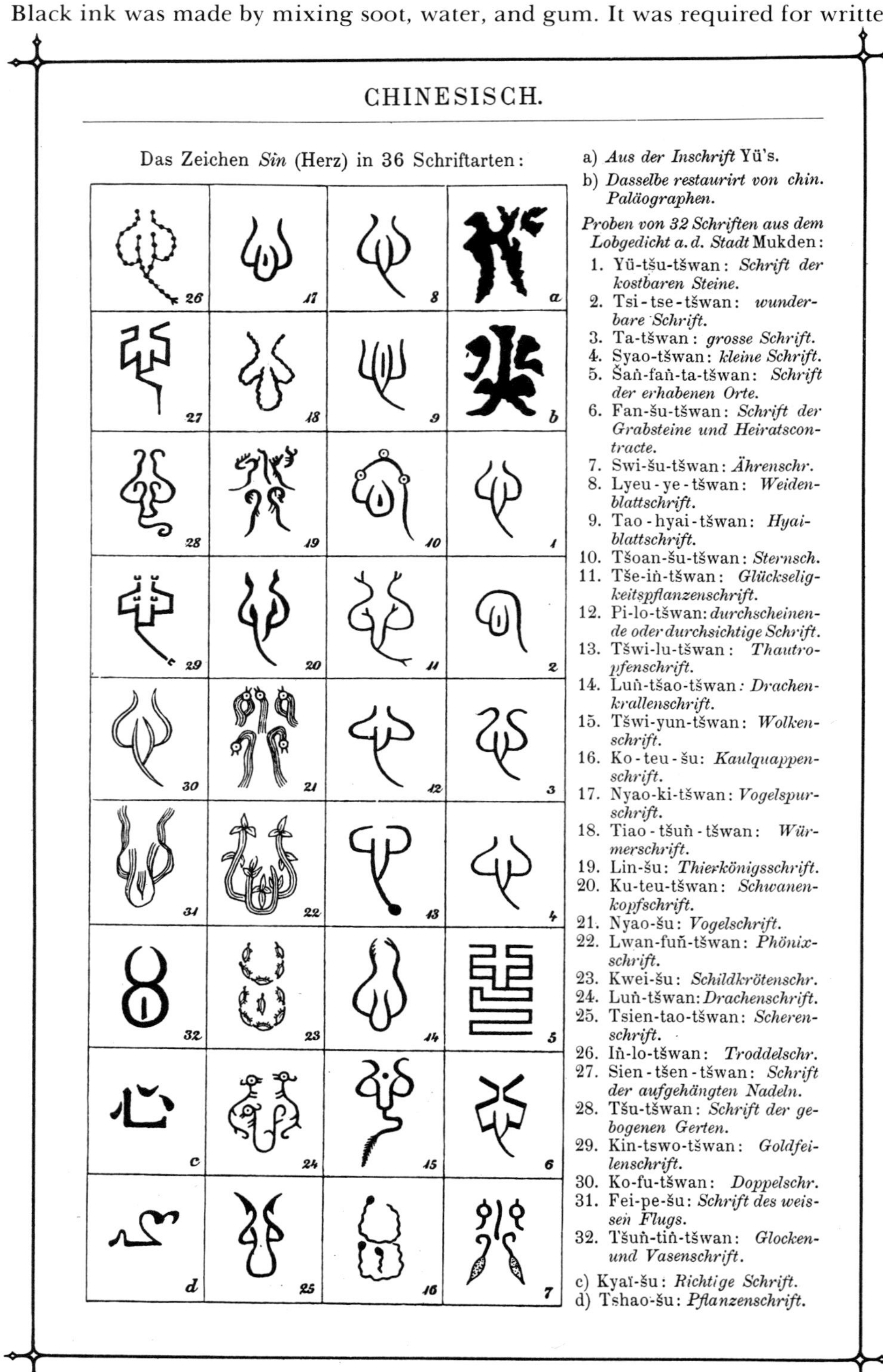

CHINESE

The sign Sin (heart) in 36 forms
a) From an inscription.
b) The same, restored by Chinese paleographers.
1. Script of the precious stones.
2. Wonderful script.
3. Large script.
4. Small script.
5. Script of the noble places.
6. Script of the tombstones and marriage contracts.
7. Corn-ear script.
8. Meadow-grass script.
9. Hyai-leaf script.
10. Star script.
11. Good-luck-plant script.
12. Luminous or transparent script.
13. Dewdrop script.
14. Dragon's-claw script.
15. Cloud script.
16. Tadpole script.
17. Bird-track script.
18. Worm script.
19. King-of-the-beasts script.
20. Swan's-head script.
21. Bird script.
22. Phoenix script.
23. Turtle script.
24. Dragon script.
25. Shear script.
26. Sword-knot script.
27. Script of the hung needles.
28. Script of the bent canes.
29. Gold-file script.
30. Double script.
31. Script of the white flight.
32. Bell- and vase-script.
c) Correct script.
d) Plant script.

The Chinese symbol "Sin" (heart) in 36 different forms.

texts. Red ink was made in the same way, with the addition of oxides. The Egyptian scribes used it to make titles, headlines, and the beginnings of new chapters stand out.

The writing instrument also "grew" on the shores of the Nile. Thin reeds were chewed and reshaped into brushes at one end. Their cellular structure, like a sponge, soaked up as much ink as was needed for a line of writing. Later a small bamboo staff was used for fine writing: the first writing tube was born—with a cut and split point in the shape of the later quill pens. The cut surfaces were carefully smoothed with a whetstone. Among the equipment of an official scribe there were a bag for pigment powder, a brush container, and a long, thin wooden palette with two wells for mixing the red and black ink with water.

Over the millennia, papyrus was to remain a desirable and popular material for writing in Egypt, during the early Greek and Roman empires, and even into the early Middle Ages.

As had the Sumerians and Egyptians, the Chinese turned their word pictures into picture writing. The Chinese language has one special characteristic: it is spoken almost exclusively in one-syllable words. Naturally the potential for expressing a manifold language in single syllables is limited. Thus the Chinese had to pronounce the 400 to 900 words of their language in four different tones to give words with similar sounds different meanings, which are understandable only in context. An alphabet was not capable of reproducing a small number of short, similarly written words with up to forty different definitions. In China, therefore, a system of written symbols composed of sounds and ideographs had to be developed, of which two or three thousand had to be memorized by scribes for everyday use. The complete Chinese writing system grew over the millennia to more than 50,000 symbols. This huge number of written symbols is hardly ever mastered completely, even by university professors.

كتاب القرائت و المنشات عثمانى

The Arabic script had its origin in the script of the Phoenicians and Egyptians. This is the title of an Osmanli reader and exercise book.

For ages, writing has been done in China using a brush and drawing ink. The inventor of the drawing ink, which unlike ordinary ink dries to be unfading and waterproof, was Tien-Chen (ca. 2697-2597 B.C.). Its production process was kept secret for many years. We know only that soot and oil were mixed with lime and kneaded. After a long process, the mass was pressed into wooden forms and dried. This ink-stone was later rubbed with water. Finely cut bamboo mats were used for writing, until the former palace guard and eunuch, Tsai-Lun (also called King Chong) invented the making of paper in A.D. 105. The invention attributed to him is based on the ancient production of felt, in which animal hairs are worked into sheets. To make paper, Tsai-Lun had bamboo shoots, rags, and hemp pounded and cooked. Out of the resulting mixture of fibers he created the sheets of paper with fine sieves and dried them in the sun on high walls.

The Sultan's signature: The calligraphic symbol of a hand with five outstretched fingers means: Abdul-Medzid, son of Mahmud the Ever-Victorious.

The brush, made of rabbit hairs inserted into a bamboo tube to form a pointed tip, was of great importance. Running from top to bottom and right to left, the writing process itself was somewhat of a ritual:

"First one shall quickly bend the brush, then quickly draw downward, as an eagle plunges down on a large bird. Without stopping to think, one must let the brush glide naturally. It must happen by itself and should never be changed. When one draws the feet (thickness at the bottom when ending the stroke), that must happen playfully, just as fishes move in water. The brush should dance like a cloud over a beautiful mountain landscape. It should either roll or spread out, lightly or heavily. When one considers these basic principles, everything will be clearly recognizable." (Li-sse, ca. 215 B.C.).

Almost a thousand years passed before paper made its way to the West. The Arabs learned the secret of making this new writing material from Chinese prisoners in Samarkand in the middle of the eighth century, and set up their own paper industry. The Caliph Haroun al-Rashid, known from the *Thousand and One Nights,* ordered his chancelleries to use paper instead of parchment and papyrus at the beginning of the ninth century. Paper reached Europe via the middle east.

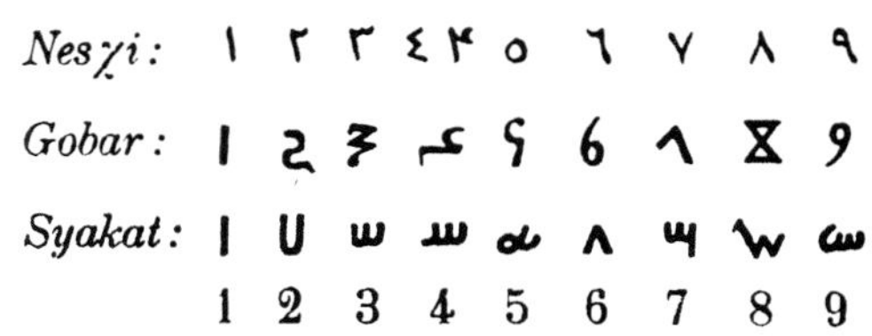

After learning the Indian numerals and decimal system, the Arabs developed the "Arabic numerals" known to us today.

The Geis Mill, the first German paper mill, built by Ulman Stromer, near Nürnberg, in 1390. As early as 1200, while Spain was occupied by the Arabs, the first European paper mill is said to have been opened in the city of Seville, home of half a million people at that time. Paper reached Germany by way of Italy. The Nürnberg resident Ulman Stromer, born in 1329, first dealt in Italian paper, until Italian specialists helped him to build the first German paper mill on the Pegnitz. "I, Ulman Stromer, was the first to make paper, at St. John's . . ."

For papermaking in China, bamboo shoots were cut and soaked. (1)

After pounding, the fibrous pulp was strained through fine-mesh sieves. (2)

Drying sheets of paper on the wall. (3)

We can thank the Phoenicians for the basic structure of the alphabet. These ancient trading people colonized the coasts of the Mediterranean Sea beginning in the tenth century. The practical Phoenicians took the basis of the system of writing from the Babylonian cuneiform. Instead of one symbol per syllable, they used one symbol per sound in their writing. Their script did completely without symbols that resembled pictographs. Instead the Phoenicians developed clear, easily written letters. From the Phoenician script, the Aramaic people (Aramaic is the language of Jesus) developed the Hebrew alphabet. The Arabic script is also based on the Phoenician, as are the hieratic and demotic alphabets of the Egyptians. In the Moslem state, writing was regarded as serving God. The use of the Egyptian tube pen on leather, papyrus, parchment and later on paper developed from a mere means of communication into a calligraphic ritual. The ninety-nine names of Allah were written from left to right in brown or black ink, or even, to show high honor, in gold or silver ink on blue or black paper, and ornately decorated. Arabic culture was closely linked to Arabic writing. According to tradition, Allah dictated the ninety-six books of the Koran to His prophet Mohammed around A.D. 630. In the Holy War, the Arabic script was spread through North Africa, Spain, and Mesopotamia, and across Persia to India. It became an international language for art and science.

The oldest Greek inscription known to us is on a large earthenware vase of the eighth century B.C. It shows obvious resemblances to the Phoenician alphabet. To be sure, the Greeks were not satisfied with writing the twenty-two sound combinations. Around the middle of the ninth century B.C. they simply turned a few of the Phoenician consonant signs into vowel signs. Thus the alphabet was born. Like the Phoenicians, the Greeks originally wrote one line from right to left, the next from left to right, and so on. It was only in the fourth century B.C. that writing from left to right prevailed.

The Greeks valued a beautiful script, especially when it was to be carved into stone and passed on to posterity. All twenty-four letters of the Greek alphabet fit into a rectangular space and were arranged in a line with regular spacing. In ancient Athens, engraved stones were placed in the city's important squares to proclaim governmental decrees or announce coming events. The chiseled "Kapitalis" of large letters was followed in the ninth century by the simpler writing of small letters, the Greek minuscule.

The minuscule was the writing of the poets, philosophers and scientists. Aeschylus, Homer and Archimedes, and Pythagoras, too, wrote it on wooden

1

2

3

tablets that were covered with wax or white paint. On their deltion or grammation they carved their verses or calculations with a stylus of bone or metal, the stilo. As in ancient Egypt, the writing instruments were rounded or flattened at the far end so that errors in the soft wax could be smoothed out immediately. Cut grooves prevented the fingers from slipping. Tied together with cords into small books, two (diptycha) or three (triptycha) wax tablets could be carried on one's belt as the sign of a literate man.

In addition, the reed pen, the calamus, was used for writing on papyrus, which must have been imported from Egypt, and on prepared animal skins. Hundreds of thousands of precious rolls of writing are said to have filled the Hellenic libraries in Alexandria and Pergamon.

The Phoenicians developed the basis of our present-day alphabet. The original forms of our present-day letters are easily recognized.

The Roman conquerors of the Mediterranean area took possession of two scripts for their empire: the Etruscan alphabet, which was based on the Phoenician, and the Greek alphabet. The letters A, B, E, Z, I, K, M, N, O, T, X and Y remained unchanged. The other letters were adapted from the Greek. While the Greek letters still had strokes of equal width, the Romans added their "Capitalis monumentalis" serifs (transverse end lines) and made the lines appear thick or thin, as had been done with the brush or reed pen. Like the Greeks, the Romans used capitals for inscriptions on stone monuments and public buildings. The merchant, the scholar or the Roman lady, though, used a flowing written script that they could write on papyrus with a brush, reed pen, or quill, or scratch on a wax tablet with a stylus. From the "Capitalis quadrata" used in documents there developed a finer script, written with a sharp quill and better suited to longer texts: the "Capitalis rustica."

The Romans also took their writing materials from the Greeks: the papyrus, sensitive to dampness, called "charta," the robust parchment "membrana," and the wax tablet. For letters and short notes the receiver was sent a small tablet, the "tabella," which was written on again and sent back after the writing had been erased.

The educated person always carried a stylus with him in a container. In case of danger, it could be used as a weapon: Caesar wounded Casca, one of his murderers, with a metal "stilus."

To protect wax tablets from falsification, they were often made in triplicate and locked up. Copies were called "exemplare." There was even a shorthand in ancient Rome. Speed writers took down politicians' speeches in the Forum Romanorum or the Capitol in the so-called "notae vulgaris," transposed them into regular script and circulated them.

Letters from the mysterious Germanic north, which reached sunny Rome via a well-organized postal service, reported on the strange customs of the "barbarians." "They celebrate in old songs, which are their only means of passing on information. The secrets of letters are unknown to the men as well as the women," the Roman historian, Tacitus, wrote. The writing of the Teutons was limited to the mysterious runes (the very name means "secrets, secret information"), which are still to be seen today on rocks and ruins in Scandinavia and England. The God Wotan, according to Germanic tradition, is said to have written them personally. Runes were carved in stone and in wood. Tablets of beechwood were called "bok," the source of the English word "book." The beechwood as writing material and the staff as stylus were the origin of the German word "Buchstabe" (letter). Wood played an important role in transmitting information in the heavily forested Germany. Messengers were given a tally stick with carved signs, the "Kerbholz," to carry the news. When someone had something "on the Kerbholz," it meant he was the possessor and transmitter of important, hitherto unknown information.

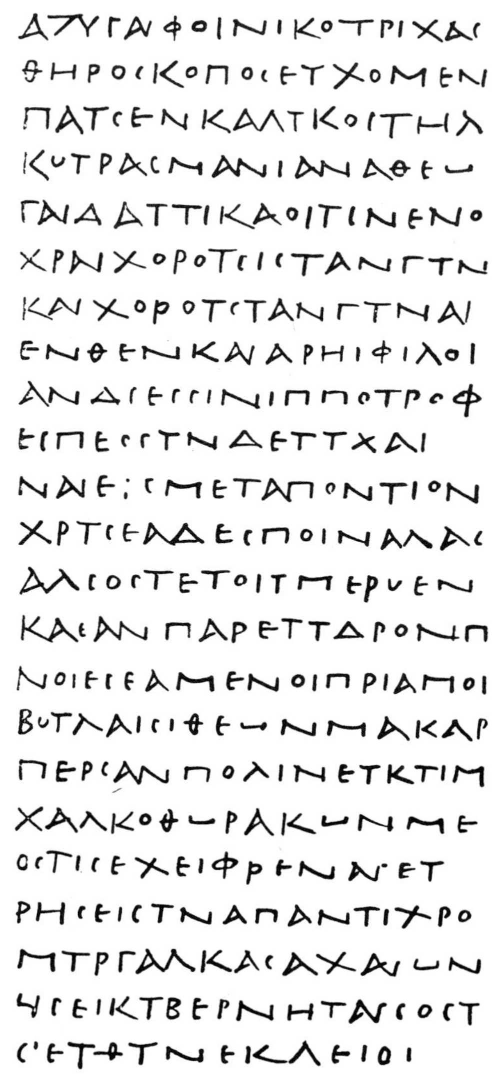

Greek book and letter writing.

Roman writing pen of sheet bronze of silver.

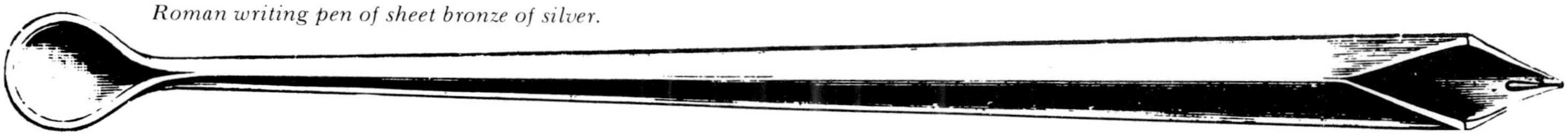

Scriptorium

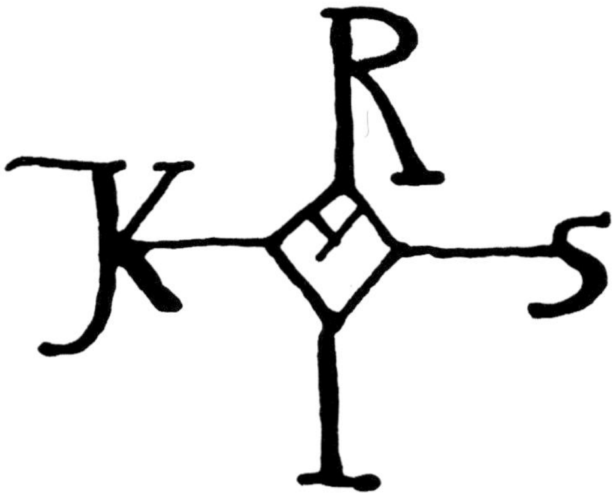

With the decline of the Roman Empire, the people's ability to write was largely lost. Even progressive rulers like Charlemagne added only the last two strokes of the diamond-shaped O by hand.

With the downfall of the Roman Empire, whose inhabitants were almost all literate, the reading and writing ability of the people was lost to a great extent. Only in the quiet writing rooms of the Italian, German, Irish and English cloisters did the cultural techniques of reading and writing live on. Scribes among the monks devoted their time to copying the bible and theological texts and reproducing the writings of ancient authors. Gradually the Roman current script developed into a writing of strongly rounded capital letters: the unzial script. The influence of English cloisters led to a script of small letters, which likewise originated in the classic Roman script: the half-unzial. There were also many regional mixtures including the west Gothic minuscule, the Merovingian book script, and the southern Italian Benevantic script. The rest of the world got along without letters and numbers. Theodoric, the King of the East Goths (454-526), never learned to write four letters in his entire reign. For his signature he had to use a monogram stamp of golden sheet metal. Even Charlemagne, in whose reign a new handwriting, the Carolingian minuscule, developed, made very little progress in writing, although he kept "a writing tablet and a book under his pillow." Charles, who was determined to bring art and scholarship back to life, had his signature "Carolus" written by a scribe. He wrote only the last two strokes of the diamond-shaped "O" himself.

Whereas the Carolingian minuscule was still based on the Roman architectural style, the Gothic minuscule showed a clear relationship with the developing Gothic style. Its appearance is reminiscent of a woven design and was thus called "Textura." As the writing of the monks, it was also the liturgical script, used above all for large missals. It was not surprising that Gutenberg used this script as the example for his first type cast in lead. He was eager to make printing look as much as possible like handwriting.

The Roman writing instruments retained their importance in the Middle Ages. Mundane scribes in chancelleries and construction offices, scholars, and merchants documented their notes, calculations, drawings, and short messages on wax and slate tablets, as in antiquity. They made use of the advantage that their distant ancestors had valued: what was written could be erased easily and economically at any time.

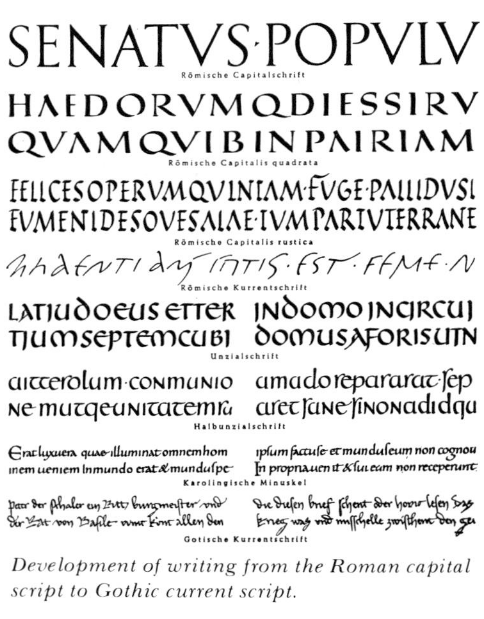

Development of writing from the Roman capital script to Gothic current script.

For over a century the cloisters remained the centers of writing. Writing ranked second only to preaching as work that was pleasing to God, especially among the Benedictine monks, whose rules required handwork in the service of the Lord. In the dim light—thin parchment was the "windowpane"—of the "Scriptorium," the writing room of the cloister, letter after letter was fervently copied from an original onto a scroll. It took months and years until a single copy of the bible was finished. Papyrus strips were usually glued together into thick scrolls. Parchment could be inscribed on both sides and was better suited to small formats. The monks went back to Roman examples by binding parchment pages between protective wooden or leather covers, thus creating the book for the space-saving preservation of written works. The bound "Codex" had another practical advantage: unlike the papyrus roll, which had to be unrolled to find a certain place in the text, one could find the desired spot in the text quickly by opening the book to the right page.

The use of classic papyrus had a disadvantage. This writing material, coming from Egypt, was not always available in the desired quantity. Home-grown animal skins, from sheep, goats and calves, were easier to obtain. Their preparation, to be sure, was very time-consuming. In the production of parchment, unlike that of leather, the animal skins were not tanned. Repeated washing in lime water brought on the corrosive process. The hairs were loosened, the fat was removed. A crescent-shaped scraping iron was used to remove bits of flesh and hair and smooth uneven spots. Then the skin was stretched in a wooden frame and thoroughly smoothed. Then the raw product was scraped on both sides with the scraping iron. After chalk was poured on to whiten it, the skin had to dry in the frame for a long time. The finished parchment now had a somewhat rougher outer (hair) side and a better, because it was smoother, white inner (flesh) side, called "album"—a concept still used today for books in which photos or postage stamps are preserved.

Whoever wanted particular luxury, and could afford it, chose the rare virgin parchment made of the skin of unborn lambs for his correspondence. Parchment was reserved for important writings such as deeds, documents of kings, nobles and high clergy, and copies of missals and bibles. Parchment was valued not only for its durability but also because writing could be scratched off of it and the valuable material used again, though this did not increase legibility.

Numerous pictorial and written documents from the Middle Ages report of the work processes that were necessary to produce a book. The contemporary portrayals offered on the following pages are especially striking.

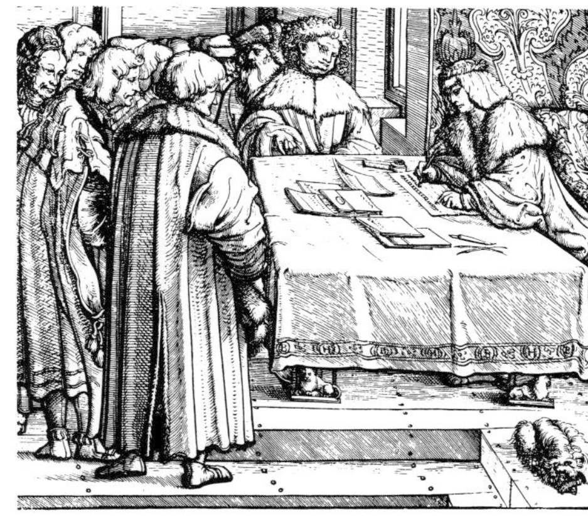

Parchment, the royal writing material at the court of the later Emperor Maximilian I. Woodcut from 1514-16.

Scribes of the High Middle Ages, from the Evangel Book of Henry III.

The 42-line Gutenberg Bible was printed in Textura in the 15th century.

Calligraphy—the Art of Beautiful Writing

In our times calligraphy, the art of beautiful writing, has been rediscovered. The word calligraphy comes from Greek (kallos = beautiful, graphein = writing). Calligraphy refers to beautiful penmanship, as opposed to everyday writing, its purpose being to form the letters not only clearly but as beautifully as possible.

In the electronic age we have long since become accustomed to carrying on business and private correspondence with the help of typewriters, computers, and other machines. Information appears on one's monitor screen via fax without even being put to paper. The sender adds only his signature to send a little bit of his personality. The farther technical progress moves, the greater is the desire for individual expression, and the stronger the need to detach oneself from the uniformity of the printout.

We find the forerunners of all calligraphy in the Chinese and Japanese writing arts and on the Egyptian papyrus rolls, which date back to the second millennium B.C. The script and material had a strong influence on the late-antique manuscripts as well as those from Byzantium. Also ranking as masterpieces of calligraphy are the valuable manuscripts of the Middle Ages with their marginal decorations and illustrations. In the fourteenth century written manuscripts were being displaced by printed works, thanks to the invention of movable type by Johannes Gensfleisch, known as Gutenberg. Nevertheless writing schools were established in St. Gallen, Fulda, and Regensburg, among other places, where the art of beautiful writing was still being advanced and taught. Writing masters such as Johann Neudörffer of Nürnberg developed textbooks and books of examples for writing school students and for adult self-instruction. The first book of instructional examples to appear in Germany, in 1519, included alphabets and examples of texts in the form of woodcuts or copperplates in everyday, chancellery, and Gothic styles.

Though the quill pen still allowed a somewhat expressive handwriting from the fifteenth to the eighteenth centuries, the introduction of the pointed steel pen limited individual handwriting to a thin line. The reproduction of the examples in many pattern books must have been misery for the

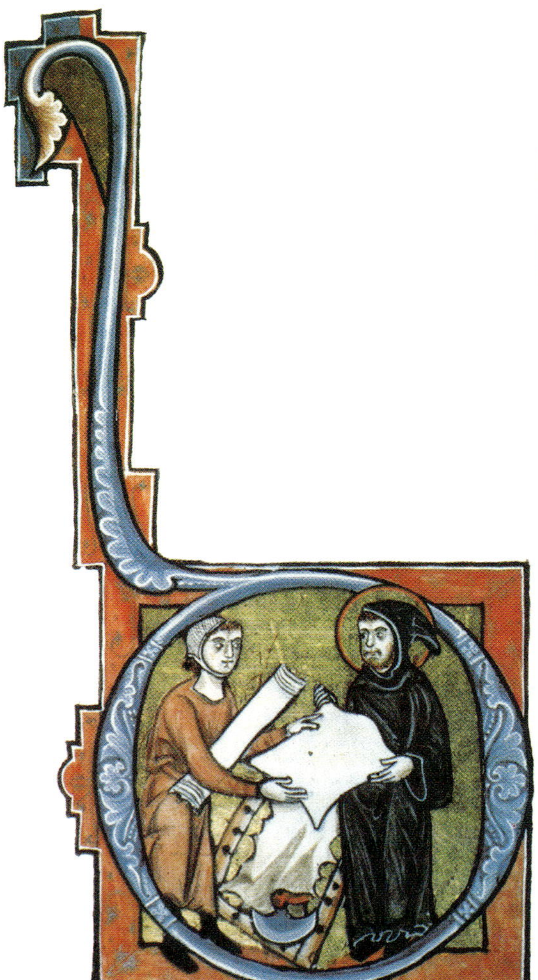

A Hamburg manuscript dating from 1255 shows how the monks produced books.

A parchment maker (a layman) delivers the desirable writing material to the monk in the black robe. In the background are the frame on which the animal skin is stretched to dry, and the semicircular scraper.

Cutting the reed pens with a penknife. At right, the surface of the parchment is made smooth or rough, depending on its intended use.

Cutting the pages and preparing the writing surface.

students. For unlike the flowingly written examples, the engravers transferred the text to the turning copper plate with a steel cutter that moved gently back and forth. The result was a beautified handwriting that could not possibly be copied with a quill or steel pen.

It was particularly the chancellery clerks who sought instruction. They had to prepare deeds, diplomas, patents of nobility and documents of many kinds, as well as often publishing patterns of their own. But soon the layman also wanted to show culture and education by writing a beautiful hand. The merchant made use of calligraphic instruction in his correspondence and record books, just as the bourgeois did in his letter-writing. For many decades following the invention of the typewriter handwriting was more and more displaced. Only a few people, such as Vienna's Rudolf Larisch around the turn of the century, Ernst Schneidler and his student Ana Aschoff, and Rudo Spermann, Walter Tiemann, and Rudolf Koch as colleagues in the Klingsporsche Schriftgiesserei in Offenbach, carried on the tradition of calligraphy in Germany. Today the display of Karl and Wilhelm Klingspor's script casting in the Klingspor Museum in Offenbach attracts interested visitors from all over the world.

In the 1980s, calligraphy experienced a renaissance, as did so many other artistic and artisan activities, as a means of very personal expression. Artists, officials, students and homemakers trying to learn the art, develop creativity, and express their personality in

Medieval scribe's workshop.

Marking the letter spaces with a sharp stylus.

The advantage of parchment: Errors can be scratched off with the penknife.

Illustrators decorated the written book with miniatures and larger paintings.

calligraphy as a hobby. In the process, a wide variety of results emerge. They range from word pictures, greeting cards, handwritten documents, and mottoes to hang on the wall, to complete, handwritten books.

The layman finds support in every good stationery shop. The industry has long been aware of the trend toward beautiful writing and offers a wide assortment of specially formed pens for the calligraphic hobby. "bond nib" pens are available with wide points for drawing lines. This classic type of calligraphy pen corresponds to the reed or goose-quill pen and allows the characteristic letters to be formed. Depending on the direction in which it is moved, it forms lines of varying width when letters are drawn.

"Even stroke" pens have a point with a special cut that allows lines of equal thickness to be drawn, independent of the direction of writing.

Drawing and writing pens have a fine hand-prepared osmium-iridium point at the end of their nib. This writing and drawing tip allows quick, fine, and even strokes and lines to be drawn without slowing the pen.

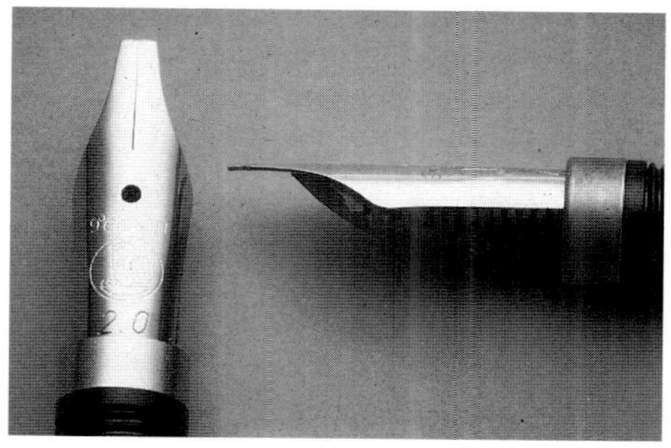

The band nib produces the characteristic varying widths of antiqua script.

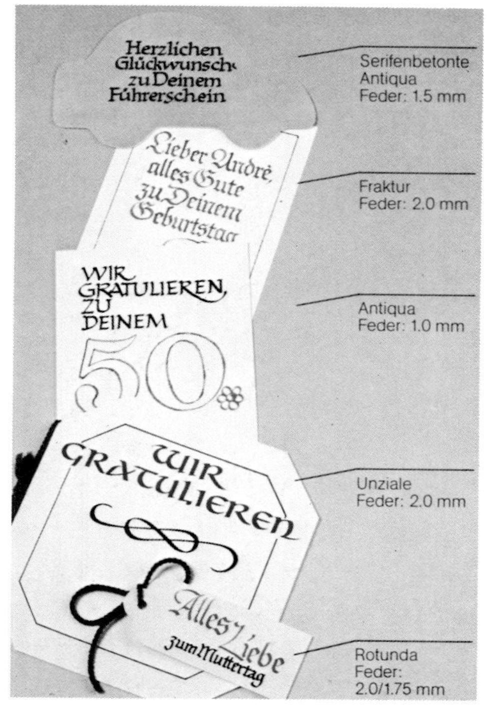

The special cut of the even-stroke pen guarantees a stroke that is always the same width, independent of the writing direction. It is used particularly for block printing.

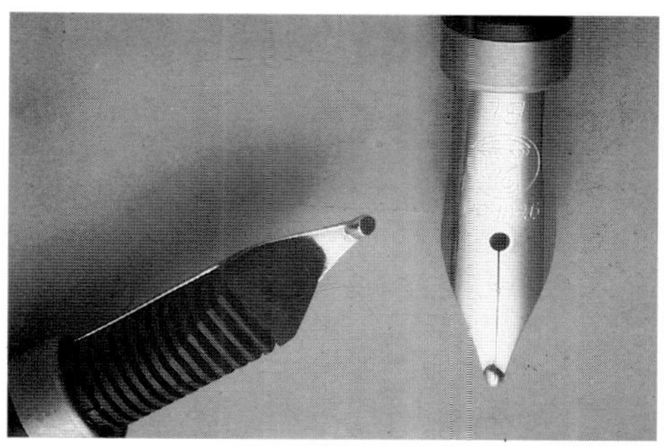

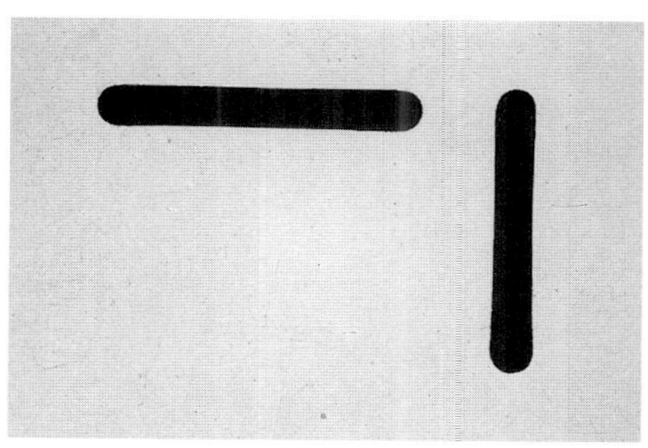

△
The examples show what effects can be made for decorative writing of very ordinary and also unusual greeting cards.

Framed quotations and dedications make impressive personal gifts.

A modern calligraphic fountain pen with changeable nibs.

The Ink

"People, buy ink!"
(Street vendors' cry)

The medieval scribe's working environment. The monk Hideberts aims at a mouse: "Cursed mouse, often enough you make me angry—may God destroy you!" The desk has cutouts for two inkhorns for black and red ink and two goose-quills.

"Aleppo oak-gall 42 portions and Holland madder 3 portions are diluted with enough warm water to make 120 portions of the liquid. After filtering, one adds one and one-fifth portions indigo solution, five and one-fifth portions iron vitriol and two portions oak-gall iron solution."

It was January 4, 1856. In the Kingdom of Hannover, the Dresden perfume manufacturer August Leonardi was granted the patent for a truly remarkable invention: the first indelible, permanent writing ink. The Alizarin ink was a "magic ink" that flowed from the pen bluish-green but changed its color to a deep black in a few moments and was not faded by the effect of either water or light.

"It forms neither a crust on the steel pen nor a deposit in the ink bottles. It is indestructible and resists the effects of chemicals, steam and time, and never turns pale . . .," the label on the ink bottle promises, and it was in demand everywhere.

Naturally? Not at all! There were acceptable inks, in fact, before our calendar time began. The Chinese and Egyptians used finely powdered bits of soot in a mixture of bone lime, blood serum, or gum. These writing fluids could be exposed to light for long periods without fading, but had the unpleasant characteristic of being scratched off easily. For the soot particles could not penetrate far into the papyrus, parchment, or paper. The Greeks and Romans experimented with chemical solutions such as were used in tanning skins. Iron salts in tannic acid solution provided an intense black color. Philon of Byzantium (250 B.C.) already knew how to write with invisible ink: Colorless tannic acid put on the writing surface could be made visible at any time by being painted over with, or dipped in, iron salts. Tannic acid was extracted from the oak-gall. This is a spherical growth that is caused by the gall-wasp stinging the leaves of the oak tree. The oak-gall ink, named for its component of tannic acid (gall and tannin) and containing iron, was the favorite writing ink in ancient times and in the Middle Ages. The monks in the cloister writing rooms, being its major users, were naturally very concerned about the durability of the ink they used. Their bibles, testaments, and scientific texts, written by hand and carefully decorated in years of work, had an incalculable value and were thus made to last for centuries. Along with a knowledge of writing and reading, the scribe had to know at least a little ink chemistry. Almost every cloister had its own formula for ink.

A written page with ink decay, the result of iron-gall ink.

In 1100, the monk Theophilus recorded his instructions for the preparation of thorn ink, as was customary from the sixth to the thirteenth century: "Thorn twigs of the blackthorn should be collected just before they leaf out in April or May and later be beaten with a hammer. The rind that comes off must be covered with water, left in it for three days, and finally boiled off. The procedure is to be repeated several times until a red-brown, thick, liquid mass is left. This should be mixed with wine, put into a parchment bag to dry, and dissolved in warm wine for use."

To make the ink darker, one stuck a glowing iron rod into the liquid and added "atrementum" (candle soot or copper vitriol) to it. Thorn inks were transparent and similar to lacquer. They differ from the black inks in their high resistance to light and water.

As containers the monks used open cattle horns, into which one simply poured the amount of ink needed for a few minutes' writing. The so-called "spike" was fastened to the writing desk with a nail or stuck into a cutout space. Red inks (rubrica), used for emphasizing initials or titles, gave rise to the German verb "rubrizieren," to make red. The brick-red ink made with "minium" (minium, or red lead) gave the present-day term "miniature."

Among the costliest of all colors was purple. It was derived from the secretions or purple snails from the Mediterranean by removing the glands from their living bodies. The gain from this gruesome procedure was meager. To produce only 1.2

kilograms of crystalline purple dyestuff some ten thousand snails had to give their lives. Because of their particular qualities, the iron-gall ink maintained its position over all competition through the centuries. Often soot, gum, or vinegar was added to it, as well as red wine yeast for brown coloring, minium for a vivid red, and now and then a bit of wormwood, to discourage the mice in the dark, bleak cloister chambers from chewing on the manuscripts. A lack of basic knowledge among the inhabitants of the cloisters, more than once resulted in the iron-gall ink eating up the manuscripts over a period of years. An overdose of acid was usually the cause.

Outside the strict and isolated atmosphere of the cloisters, there was also a "worldly" use of writing fluid. Wandering peddlers offered not only objects such as brushes, combs and pots, but also ink carried in small barrels slung over their shoulders or in larger wooden containers carried on beasts of burden. Calling or singing loudly, they attracted chancellery employees, merchants' apprentices or citizens' daughters to the street and, using measuring cups and funnels, poured their homemade ink into the containers the customers held out to them.

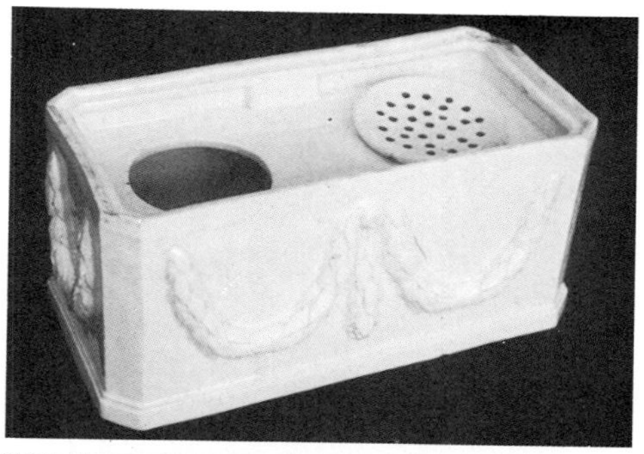

Artistically made of wood, tin, burnt clay or porcelain, worked inkstands decorated desks over the centuries. Along with the ink bottle, they also included the box of blotting sand for drying the writing.

An elegant brass Soennecken inkstand of 1890.

1

2

3

4

English vendor's cry of the early 18th century (1)

Selling ink, circa 1821 (2)

"Buy ink!." Vienna, circa 1775 (3)

Italian ink vendors at the beginning of the 17th century (4).

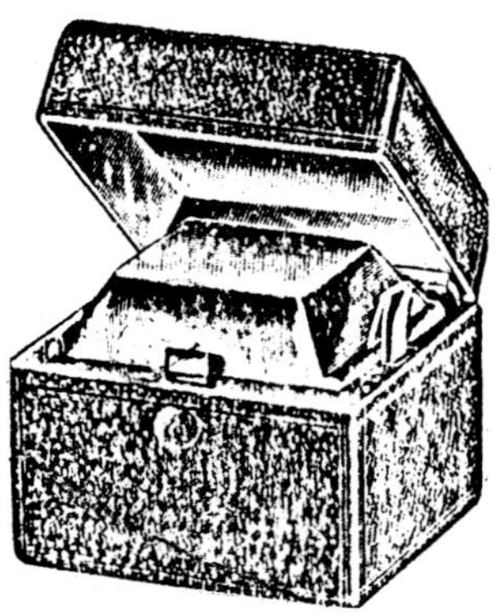

Pocket inkwells were part of the traveler' equipment from 1860 to about 1910. The sprun; cork lid could be opened by pressing with th thumb.

All inks had one thing in common: They proved to be faulty when used. Because the tended to separate, the dark liquids derived from cooking iron salts and tannic acid could only be used for writing when mixed with water-soluble thickeners—usually gum. To be sure, the color pigments no longer separated from the liquids, and the ink adhered to the paper better, but at the same time, considerable difficulties appeared. If only a little bit of the water in the opened bottle evaporated, the ink quickly became slimy. Crusts formed on the pen, which crumbled and broke off under the effect of the acid, and resulted in high pen-point consumption. And not least, the inks of past centuries spoiled quickly—they simply became moldy.

Along with the classic iron-gall inks, logwood ink also flourished briefly at the end of the past century and the beginning of ours. It was invented in 1842 by the German chemist Runge. Its main constituent was logwood extract, derived from the wood of a tree (Campeche wood) that grew only in Central America. This reddish dyestuff oxidized with chromic acid phosphate to a dark purple before it faded, producing an inexpensive ink pigment.

August Leonardi, of the small kingdom of Hannover, solved several ink problems at one stroke. He used the main constituents of tannic acid and iron salts not as a solution, but separated from each other. In the process, he utilized the principle of invisible ink developed by Philon of Byzantium more than 2000 years before. Leonardi put the colorless tannic acid and iron salt mixture on paper. When it dried the iron salts oxidized and took on color. To write with a colorless ink is difficult at best so Leonardi first mixed it with a temporary coloring material of madder root—Alizarin—and indigo. This trick made the colorless tannic acid immediately visible on the paper, in a blue-green color. As soon as the water evaporated from the paper and the ink dried the intense black color of the iron-gall ink appeared on the paper. Thus Leonardi could dispense with gum additives. He was also the first to add an anti-bacterial wood vinegar when storing his ink. Thus a lasting and easily flowing ink was born—a necessity for the functioning of the first fountain pen, which was just coming into use.

Despite the universally used present-day inks with their aniline coloring materials ("royal blue" ink, for example, is the favorite fountain pen ink, preferred by 90% of pen users), the blue-black "Alizarin ink" (= iron-gall ink) has not disappeared and is still produced today. One of the reasons for it is surely a decree of the Imperial German Chancellery of 1888, which prescribes that documents may be filled out only with authentic, officially tested iron-gall ink. For only with it are falsifications or corrections (erasures) always provable under a quartz lamp by modern processes—which cannot be said of the colorful dyestuff inks. They disappear without a trace when they have been exposed to sunlight long enough. Letters written with royal blue ink can even be erased with an ink eraser and corrected without leaving traces, to the delight of all students.

Another ink problem, that of transporting ink in small, fragile glass bottles on one's travels could at first be dealt with only unsatisfactorily. During any journey by mail coach over bumpy country roads, the glass bottles were in danger of being

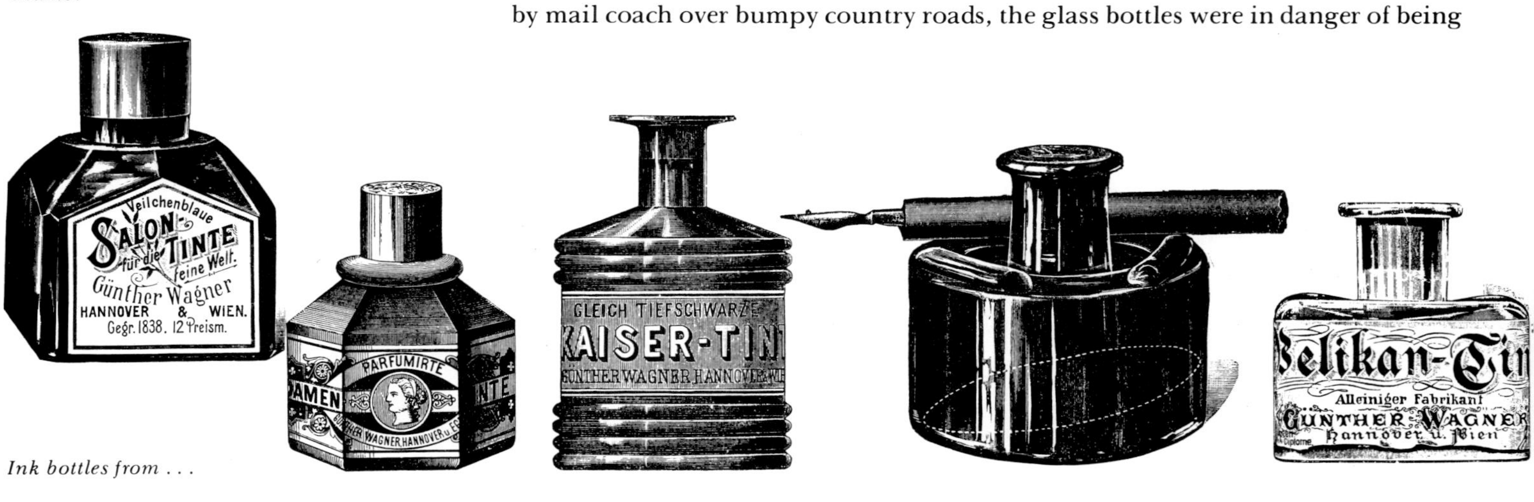

Ink bottles from . . .

broken, so that the traveler constantly had to be careful not to risk messing up the contents of his luggage with ink. The English ink industry developed a veritable cascade of ideas, by designing leakproof metal ink bottles with fastening lids and cork linings. Many types resemble gasoline-fueled cigarette lighters, while others look like miniature hats. Even opening skulls were more or less tasteful ink bottles for the nineteenth century traveler. The most widespread form was the now classic cubic ink bottle with removable neck and an aperture to hold the pen point. This basic principle was used in dozens of ways. Big heavy crystal bottles ornamented the writing desks of important personalities in commerce and government. Clever divisions even allowed the storage and use of several different shades of color in the same bottle. Filling the bottles was done economically and reasonably from liter or half-liter bottles.

Advertising by the inventor of iron-gall ink, from 1881.

Ink Production around the Turn of the Century

The drawings duplicated here are excerpts from an advertising brochure published around 1901. Along with interesting details of the manufacturing process, they include specific advice as to the uses of the various types of ink.

Apparate zum Entziehen der Gerbstoffe aus den Galläpfeln.

Eisengallus-Kopiertinten.

Die Eisengallus-Kopiertinten können sowohl für die Bücher als auch für Kopierzwecke Verwendung finden. Sie liefern ebenfalls unvergängliche Schriftzüge und geben kurze Zeit nach der Niederschrift zwei bis drei Kopien. Zum Kopieren auf der Kopierpresse ist die Eisengallus-Kopiertinte die geeignetste Tinte. Als beste Eisengallus-Kopiertinte empfehle ich die "Pelikan"-Tinte 5001.

Apparat zum Eindampfen des Galläpfel-Auszuges.

Blauholz-Kopiertinten.

Die starke Blauholz-Kopiertinte ist eine weit verbreitete und für Geschäftsbriefe sehr geschätzte Tinte. Sie liefert drei und mehr gute Kopien und bleibt monatelang kopierbar. Für die Kopiermaschine ist die Blauholz-Kopiertinte unentbehrlich. Als beste Blauholz-Kopiertinte empfehle ich die "Pelikan"-Tinte 3001.

Das Kochen der Tinten.

Das Kopieren.

Die Kopiertinten sind so zusammengesetzt, daß die Schrift nach dem Trocknen eine gewisse Zeit löslich bleibt. Sobald die Schrift mit feuchtem Kopierpapier in Berührung kommt, löst sich die Tinte. Je größer die Kopierfähigkeit eines Schriftstückes sein soll, um so leichter muß die Löslichkeit der Tinte sein. Es ist begreiflich, daß eine starke Kopiertinte, die 5 bis 6 Kopien gibt und noch nach Monaten kopiert, für die Bücher nicht benutzt werden kann.
Bei Kopiertinten, die als Buchtinte verwendbar sind, ist der Grad der Löslichkeit geringer, da-

Ein Lagerraum für unfiltrierte Tinten.

Das Kopieren.

mit die Schrift in den Büchern nicht verschmiert. Als natürliche Folge davon behalten mit einer Buch- und Kopiertinte hergestellte Schriftstücke ihre Kopierfähigkeit nur kürzere Zeit bei. Beim Kopieren ist darauf zu achten, daß das Kopierpapier gleichmäßig durchfeuchtet wird. Ob das Kopierpapier mehr oder weniger feucht sein muß, hängt von dem Alter des Schriftstückes und der verwendeten Tintensorte ab. Starke Blauholz-Kopiertinten kopieren leichter als Eisengallus-Kopiertinten. Das Papier beansprucht deshalb entsprechend weniger Feuchtigkeit.

Ein weiterer Lagerraum. Die Vorratsfässer fassen etwa 300000 Liter.

Mehrere Kopien.

Sollen von einem Schriftstück mehrere Kopien genommen werden und bedient man sich des Kopierbuches, so ist es zweckmäßig, die Kopien auf einmal zu nehmen. Die erforderliche Anzahl Seitenblätter wird übereinander gelegt, und bei längerem Pressen etwas feuchter als gewöhnlich, kopiert. Bei dem Verfahren, die Kopien nacheinander abzuziehen, tritt leicht der Übelstand auf, daß bei der ersten Kopie zuviel Kopierkraft durch zu starkes oder zu langes Pressen verbraucht wird. Bei Kopiermaschinen kann das Nacheinanderkopieren ohne Gefahr für die Ausgiebigkeit geschehen.

Das Abfüllen der Tinten.

Das Schreibpapier.

Einen großen Einfluß auf die Tinte, ihre Schreibweise und ihre Kopierfähigkeit übt das zur Verwendung gelangende Papier aus. Befriedigt eine Tinte nicht, so muß festgestellt werden, ob der Grund dafür in der Tinte oder in dem Papier zu suchen ist. Ist das Papier ungenügend geleimt, so saugt es die Tinte stark auf. Die Folge ist, daß die Schriftzüge verlaufen und durchschlagen. Schlecht geleimte Papiere sind für Kopiertinten besonders nachteilig. Sie beeinträchtigen die Kopierfähigkeit sehr. Oft zeigen auch die bei der Papierfabrikation verwandten Chemikalien ungünstige Einwirkungen auf die Tinte.

Modern ink bottle.

. . . the turn of the century.

Pelikan

Pelikan
1910
Pelikan
1920
Pelikan
1925
Pelikan
1938
Pelikan
1952–heute / 1952–today / 1952–hoy

From the family coat of arms to a worldwide trade mark.

Using pen and paper was a tradition in the founding family of the Pelikan firm. The father of the Hornemann family, a painter and copper engraver, owned a stationery shop and also held a position as drawing teacher for the children at the English court in Hannover. At that time the best artists' colors came from France and England, and the royal family of Hannover was personally linked to the latter. But this did not stop officialdom from putting heavy duties on all British colors. While one of the Hornemann sons made himself a name as an artist, the second, Carl, decided to study chemistry and start his own color production in an old farmhouse outside the gates of Hannover, in order to avoid the high costs of transport and customs duty. He was able to equal and even exceed the quality of the imported colors. His ability led to quick expansion of the new firm. By 1842, a move into a bigger building in Hannover was unavoidable. At that time Germany had 31 million inhabitants and the city of Hannover 34,000.

In 1863 the chemist Günther Wagner entered the firm as director of the factory, and later he purchased it, giving it his name. He expanded production to include writing and drawing inks and glue, and expanded the company into neighboring states. As a businessman with vision and foresight, Günther Wagner looked for a particular symbol to guarantee the dependable high quality of his products and make them stand out better than those of his competitors. He decided to use his family's old coat of arms: the pelican, standing in its nest and self-sacrificially nourishing its four young with its blood. This bird figure had already been used as a symbolic creature by the ancient Egyptians and the kings of Persia, and had been used particularly often in the religious art of the Middle Ages.

In 1878 the Pelikan symbol was one of the very first trademarks to be registered. With this concept the businessman founded one of the oldest brand names "made in Germany."

Just a few years later (in 1881), a young businessman began his career as a traveling salesman for Pelikan in Eastern Europe, Italy, and the Orient. Fritz Beindorff, later the sole proprietor, spent almost sixty years with the firm and raised it to worldwide significance through a careful development of products, extraordinary exporting success, and the opening of twelve factories in the East.

In 1901 the yearly turnover had reached a million Reichsmarks. Even in the early years Pelikan exported 35% of its production, compared with 40% today. Five years later, in 1906, a new factory on Podbielski Street, designed for the future, was begun. The epoch-making steel-and-concrete construction and the outside and inside design in Youth Style stand out to this day as an architectural monument.

Soon after the turn of the century, in an era of increasing industrialization, the production program began a planned expansion, including the addition of products used with the typewriter (typewriter ribbons, carbon paper, copying supplies), the use of which was increasing rapidly at that time.

By the 1920s, the assortment of inks was astounding. It included normal writing, special inks for bookkeeping (non-erasable), copying inks, hotel inks (easy to wash out), annotating inks (impervious to water), perfumed inks (for the elegant world), gold and silver inks, as well as inks for technical equipment. 172 different types in liquid or tablet form, colors, and bottles, were available.

It may be surprising that there were no pens, fountain or otherwise, among these products. To be sure, a Pelican fountain pen made by Thomas de la Rue, had become popular among the British public by the turn of the century, but the British Pelican fountain pen manufacturer and the German Pelikan firm did not seem to be aware of each other. In addition, production of the British Pelican pen ended long before the Pelikan firm considered producing its own pens.

In 1925 Pelikan decided to develop a fountain pen. The Hungarian engineer, Theodor Kovács, had originally developed a drawing-ink fountain pen for the firm, to be used in technical drawing. It was to become known and popular throughout the world as "Graphos." But his life work was to be the development of fountain pens that were constructed to satisfy all of their consumers' as yet unfulfilled wishes. Under the protection of numerous patents, he and his colleague Caroly Bako built the first Pelikan fountain pen, which ranked above all previously made models thanks to a series of advantages. Kovács decided in favor of the reliable piston filling system with differential screw threading. With every filling, the piston cleaned the large transparent ink reservoir, allowing one to check the ink level at any time. The soft-writing 14-karat gold pen point with its osmium-iridium tip was combined with a newly developed and patented ink regulator, which prevented the spotting caused by temperature variations. The mechanical parts were made to a tolerance of +/1/100 mm. Thanks to this precision, the Pelikan fountain penproved to be an absolutely reliable companion. Like the Volkswagen of Wolfsburg decades later, a Pelikan guaranteed absolute reliability.

The Model 100, introduced in 1929 with its typical green and black wrapper, became one of the best-selling writing instruments in Europe, South America, and Japan in the following years. By 1931 Pelikan dominated the market in its field, and expanded its offerings. The Pelikan 100 evolved into the 101, which came on the market with a white-gold or 14-karat gold case consisting of shaft and cap. The less expensive *Rappen* model, added to the line in 1932 and renamed the *Ibis* in 1935, was equipped with a simple, non-mechanical tube filling system.

An unusually beautiful item was introduced in 1935: the famous Pelikan *Toledo.* This fountain pen consisted of a sterling silver case with 24-karat gold plating and an engraved pelican symbol, following the pattern of the centuries-old goldsmithing art of the Spanish city of Toledo.

Colors were also introduced: tortoise shell brown, black, a snakeskin pattern, blue-and-red, and metallic gray. In 1935 the fountain pen was joined by the matching No. 200 mechanical pencil. To link it closely to the now-famous name, it was given the unusual brand name of *Auch-Pelikan* (Also-Pelikan).

In 1937, a year before the firm's 100th anniversary, the 100 and 101 models were improved. Equipped with a conical cap end and filling handle, plus a greater ink capacity and a larger point, the new model numbers included the letter "N," meaning new.

When World War II began, the use of gold in pen production had to give way to steel (CN = chromium-nickel), and even this ended completely in 1941.

A new chapter in the history of the Pelikan pen began in 1950. Models 400, 500, 600 and 700 continued the traditional level of quality in modern dress. For the first time, these pens bore the typical clip in the form of a pelican's bill.

The *Ibis* was succeeded in 1952 by the smaller and lower-priced Pelikan 140 and the student model 120. Fountain pens and mechanical pencils were joined for the first time in 1955 by the new Pelikan ballpoint pens, known as *Roller* 355 and 455.

In 1956 the 400 line was reworked and equipped with the upward-working filler lever and cap. The formerly irregular dark green stripes now ran regularly, and the new number was 400 NN.

The Pelikan P1 appeared in 1959. This piston-type fountain pen had a modern concealed point and a metal cap. The ink level could be seen through a small round window.

In 1960 Pelikan's marketing strategies changed with the introduction of the famous blue and silver school fountain pen with ink cartridges. The firm

Advertising design of 1953, created by the famous graphic artist Herbert Leupin.

Christmas advertisement, 1939.

concentrated on the Pelikano school pen, of which many millions were produced yearly. The P15 and P25 (1961) adult-model cartridge pens, the M/P 20, M/P 30 (1965-66), M/P 60 (1965) and M/P 100 (1968) mechanical or cartridge fountain pens surpassed the success of the Pelikan 400, 140 and 120, which were phased out in 1965, despite their first-class gold points.

As a result of changing values, an answer to the world of computers and plastics that set in as the 1970s gave way to the 1980s, the public demand for high-class fountain pens increased. Pelikan decided to take up again the manufacture of high-quality writing instruments for adults, with new offerings. The modern *Signum* line (1979) was followed in 1982 by the rebirth of the famous Pelikan 400 of the 1950s. Though technically updated in several internal parts, the *Souverän* had an unchanged, traditional form which was accepted enthusiastically by the public.

With a thoroughgoing assortment of fountain pens of various types and in various price ranges (Models 100, 150, 200, 250, 400, 600 and 800), the traditional Pelikan fountain pens presently rank among the best-selling writing instruments and inspire fountain pen fans and collectors in Europe, North America and Japan. Even the famous *Toledo 700* is a member of the new and old Pelikan pen family. And for the firm's 150th anniversary (April 28, 1988), there appeared a golden Pelikan as a numbered limited edition. In addition to writing instruments, today's Pelikan products include about 3000 items for school, office and family use.

The 400 and 450 models with pelican's-bill clip, 1950.

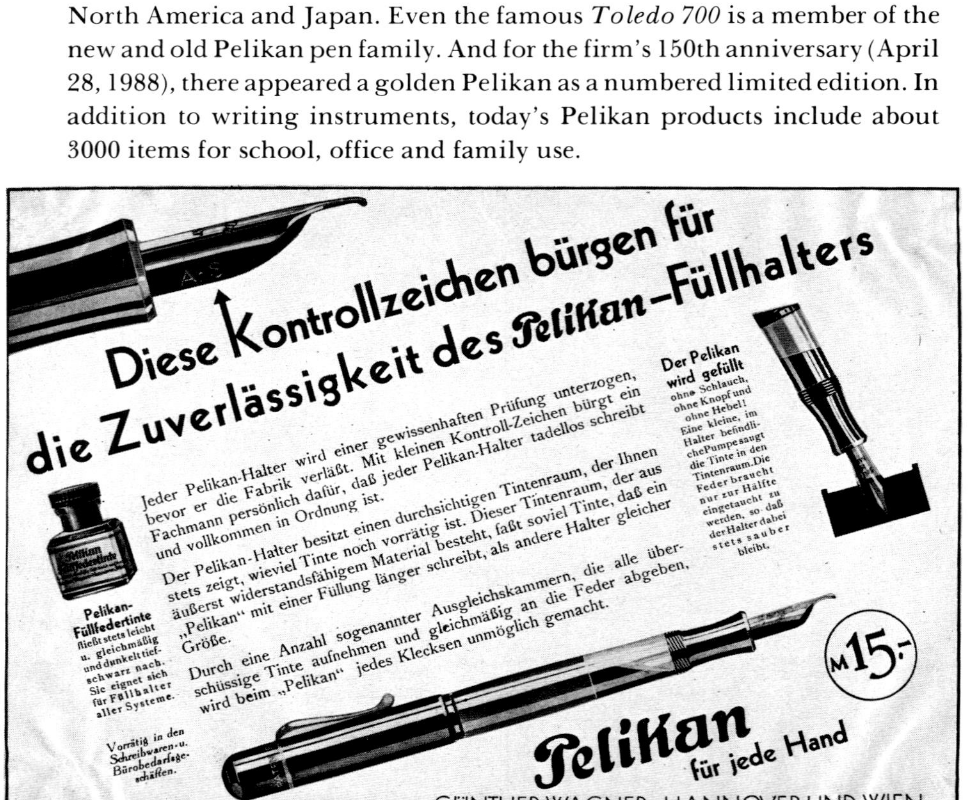

The first Pelikan fountain pen, 1930.

The elegant world, 1935.

The Pelikan 100 N with conical cap end.

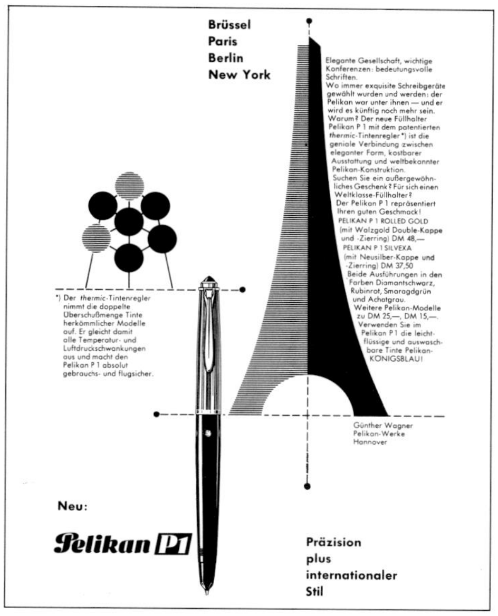

1

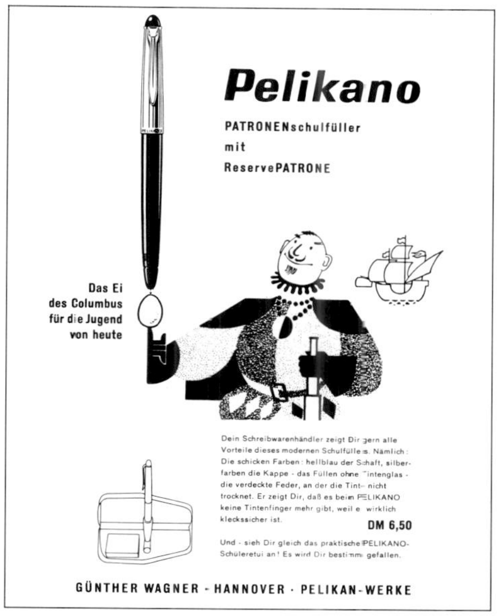

2

Modern advertising styles from the "age of the kidney-shaped table," 1959. (1)

"The Egg of Columbus" for Pelikan was the Pelikano school fountain pen, made from 1960 on. (2)

A curiosity in the firm's history: Pelikan registered a number as a trade mark, 4001.

Highest-class fountain pen decorated by a master's hand: the "Toledo."

The 400 model series—black and green, tortoise-shell brown, and with a rolled gold cap, 1950.

Advertising design, 1910. ▷

Pelikan

The Quill Pen

"Shall I write with a stylus, chisel, quill?"
Dr. Faust to Mephisto,
"Faust", by Johann Wolfgang von Goethe

The quill pen remained the preferred writing instrument over the millennia. Just above the manuscript, the penknife lies near at hand. The inkhorn, the reservoir, is attached to the side of the desk.

"But the things that one makes pens of and writes with, are goose quills, peacock feathers, foreign calamus or reeds, also copper and brass metal," reported the renowned writing master Johann Neudörffer in 1514. The most important instruments for writing on parchment and paper, from ancient Rome to modern times, remained the wide reed pen for large lettering and the quill pen. Feathers from the raven, peacock, swan or goose proved to be best suited for the sensitive surface of parchment. Whether smoothed down or left in their original condition, the natural bend of the quill fit well into the human hand. Unlike the stiff reed, the elasticity of the quill supported the elasticity of the writing rhythm. Curves and circles could be executed with more spirit and elegance without tearing the sensitive animal skin, which would have smudged the writing.

Along with ink, pen, and parchment, the penknife, a rather long, sharp blade, was the indispensable tool of those who wrote. With it one smoothed the surface, carved the disturbing filaments from the quill, scratched out mistakes and cut the point properly. Cutting the quill was an art that could be mastered by a writer only after long practice. Famous writing masters gave precise written instructions to their students, because the precision of the writing depended to a large extent on the sharp condition of the pen. "Writing students can, at least at the beginning, neither improve the pens they need nor cut new ones for themselves. The writing master must thus provide them. The writing lesson would come to grief with a big, mediocre class, and perhaps it would not even be possible to improve all the students' pens. In fact, for those who press hard, one pen will not last an hour. . ."

How highly a masterful skill at cutting a pen was valued can be seen in such German turns of phrase, still used today, as"to use a sharp pen," "a pen using member," and "quill driver." And one had to cut pens practically constantly. The relatively soft point wore out very quickly on the uneven surface, especially on the rougher outside of the parchment. Too much pressure when writing made the point spread out too far. The acid in the ink also tended to weaken the material. So the penknife had to be reached for again and again, whenever it was needed.

The beauty of the script was an important element of calligraphy, second only to its basic function of communication. Thus the writing masters not only made sure they presented excellent models in specially printed writing textbooks for self-study, but also gave advice for the correct, relaxed position of the hand when writing. Holding the pen with the thumb and index finger was not highly regarded. Three fingers—thumb, index finger and middle finger—were needed to give the writing instrument excellent position and guidance.

Often the devotion of the calligrapher went far beyond the goal of writing clear, easily legible letters and words. Raised above the necessity of writing information, notes, data and documents for chancelleries, business offices and the letter-writing public quickly and unmistakably, the calligraphers in their ivory towers joined the ranks of the elegant ornamental painters. The illustration on page 29 shows how even heroic figures could be created with designs that resembled lettering.

Goose feathers provided most of the good writing quills.

The spread of trade in the seventeenth and eighteenth centuries, the developing arts and sciences, drove the consumption of writing materials to astronomical heights. One goose supplied a mere ten to twelve good pen quills. In Germany at the beginning of the nineteenth century, about fifty million quill pens a year were consumed. The main producing nations in Europe were Russia and Poland, where geese were raised solely for the purpose of obtaining highly valued pen quills.

England imported 27 million quills a year from Petrograd. One of the major consumers, the Bank of England, used 1.5 million quill pens a year. One writer used an average of five quills a day. One can imagine how often work had to be interrupted for the careful cutting of each individual pen. The profession of the pen-cutter spread rapidly. A capable practitioner could cut up to a hundred precisely prepared pens per hour!

The London bankers, thinking in terms of thrift, looked for alternatives. In 1809 Joseph Brahman applied for a patent on a machine that was capable of cutting goose quills lengthwise and crosswise into several "nibs." About twenty of them could be made from one feather. They were finely pointed and inserted into a wooden holder. The ancestors of the pen-point and penholder had been born!

In the search for pen points that were superior to the quickly worn quill points, the glassmakers also tried their luck. German, Japanese and North American glassmakers first tried thin glass rods with sharply pointed facets. After they were dipped, a sufficient quantity of ink would remain in the facet-like grooves and slowly run down to the point. But despite certain advantages—scarcely any wear, no corrosion, the ability to write on paper, wood, leather and similar materials, a reliable supply of ink in the grooves—the glass pen was not accepted. Its writing point was too dull, and the sensitive material broke too easily. Only when Thuringian glass-blowers were able to improve the cohesion and capillary strength did the glass pen become an export success. French, American and German manufacturers put them in holders and offered the devices as ideal writing instruments for use in the office. The glass handle—a closed glass tube with an inserted spiral point—enjoyed a brief renaissance in the twentieth century, in times of economic necessity during the World Wars.

Having great difficulty cutting quill pens, the famous English chemist Priestley, the discoverer of oxygen, commissioned a toy factory in Birmingham to make a durable pen-point of metal in 1780. Though this was done with less than successful results, the metal nibs can still be admired in a Birmingham museum to this day.

Exemplary writing position (1783).

The development of calligraphy in the 17th century.

The exemplary way to hold a quill pen with three fingers. Around the edge are various penknives.

Producing Good Writing and Drawing Pens

from: The German Friend of Art, 1825

The best writing and drawing quills come from the ostrich, swan, goose and raven. The first two, which are very strong, are used primarily for writing on parchment or thick paper, just as one uses goose feathers only for ordinary writing but raven feathers for drawing and fine writing. But now drawing and fine writing are very often done with goose quills too.

The drawing of the feathers, to remove the fat and give them their characteristic hardness, is done in a very easy process. One uses either warm ashes or glowing coals. In the ashes, though, there may not be any coals, and they must also be neither too hot nor too cold. Otherwise the quill may soften too quickly and become limp instead of hard, and it would then be jagged when cut. Alternatively, the quill may not soften enough because the warmth does not penetrate all parts of the quill, and then it again becomes jagged. It is best when one works writing and drawing quills over a coal fire in the following way. At first one must be careful that the coals do not produce any flames, because then the feather held over it would burn at once. One moves the feather over the coals, and often removes it from the fire altogether, in order to feel whether it has softened thoroughly and to the same degree. If so, then one puts a woolen cloth over one's knee, takes the quill in the left hand, so that the back of it is curved downward, holds it on the cloth, and presses with the back of a wide knife against the end of the feather, cuts through it, producing a weal that can be seen through, and the outermost skin comes off. Then one presses the quill into its previous round form, by drawing it steadily through a cloth while it is still warm. Whoever prefers hard quills can draw it twice and give it two weals, but the feather must have been cooled before the second drawing.

In order to prepare good writing and drawing pens quickly, even without a coal fire, one takes a piece of coarse blotting paper, folds it over itself six to eight times. The quills to be drawn are wrapped inside it, then puts this on one's knee, holding it fast with one's flat left hand during the drawing. The feather is rolled back and forth with the right hand until it has warmed uniformly. Through this process one will also obtain nice, clean, hard quills freed from their fat.

It is an unpleasant situation when a quill becomes jagged while being cut. These "teeth," though, rarely extend through the whole thickness of the quill, but are found only on the surface. One should smooth the quill until the rough edges are eliminated. If a quill has its necessary hardness, it never develops roughness.

The quill pens from Holland and Hamburg are regarded as the best, but in Bohemia too, and especially in Siebenbürgen, one can get especially good writing and drawing pens from geese and ravens.

The necessary equipment: quill, penknife, ink bottles and a box of blotting sand, plus a bell to summon the servants.

This 1671 portrait by Gerald Dow shows the trouble that the writing teacher went to in cutting quill pens.

The right way to cut quill pens:

One held the quill in the left hand and the knife in the right one.

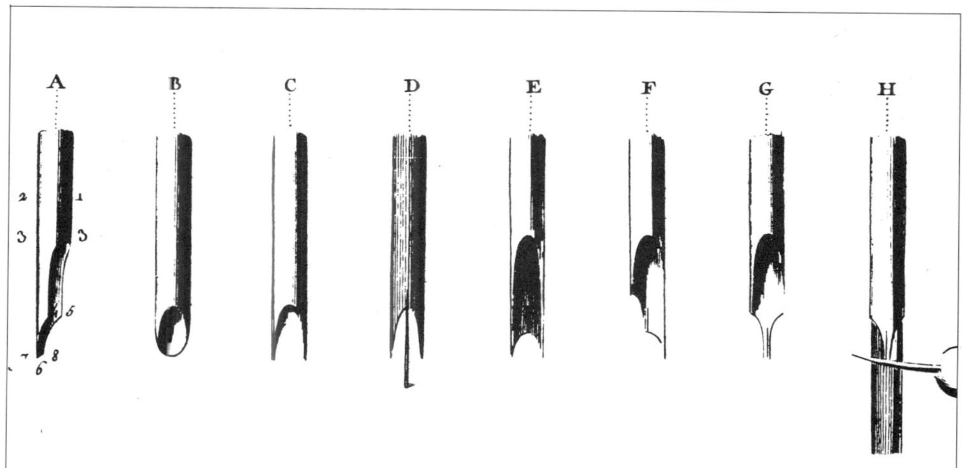

A. To make a finished quill pen,
B. one cut off the stump with a diagonal cut.
C. The same process was repeated from the opposite side.
D & E. Now one split the quill with a longitudinal cut. A vertical cut divided the subsequent shanks of the pen.
F & G. The shanks are cut out at the sides.
H. A horizontal cut shortened the point. The higher this cut was made on the conical point, the wider strokes could be made.
The elasticity depended largely on a vertical cut to divide the shanks. According to the pressure put on the pen, the writer could make fine or wide strokes.

The Steel Pen

"The devil take pens of steel!"
Eduard Mörike

A rolling advertisement for Blanzy pens.

England quickly rose to be the leading industrial nation in Europe at the beginning of the nineteenth century. The cataclysmic changes in terms of coal and iron production, inspired the thought of ceasing to import millions of goose quills and developing an idea that was not at all new. Instead of using the costly natural material like goose feathers, steel pens should be mass-produced industrially. The first step had been made by Aloys Senefelder, who invented lithography in 1798. From the spring of a pocket watch he made, by hand, a pen to inscribe lithographic stones.

The breakthrough to machine production was made by John Mitchell of Birmingham. He got the idea of stamping pen points out of thin rolled sheet steel by using a steam-driven stamping machine, such as were used in making buttons. Presumably the early metal pens, like that of Priestley, were still far from equaling the comfort of the quills that had been used for thousands of years. They had hardly any elasticity, scratched, spattered, and even made holes in the paper.

To increase the elasticity, experiments were made with various slits. The book dealer James Perry found in 1830 that a hole plus additional short slits beside the main one, or on the shoulder, could increase the steel point's flexibility. Joseph Gillot finally made a steel pen in 1831 that could compete on equal terms with the quill pen, by making the shank of the pen run parallel instead of pointed.

In 1842 seventy million pens were produced at Gillot's steel-pen factory in Birmingham. That increased to 100 million in 1843 and 180 million in 1850. How it was done was portrayed in a thorough report in the journal "Gartenlaube" in 1875:

"To produce completed products, strict work processes are carried out throughout. The pen takes on its first form in the cutting room, from thin strips of sheet steel twice as wide as the length of the pen. The pens are cut point to point, so that there is as little waste as possible. Since most processes do not require a great amount of energy, they can be carried out easily by girls. By using a perforating machine, a girl can turn out three hundred in a minute (two or more pieces at one stroke); a second, similar machine makes the holes and any side slits in the pen. Then the pens go into the stamping room, where a kind of rocker stamps various numbers as well as the name of the firm (often also that of the customer) on them. Then after the pens are heated to a glowing red, to make them soft, they are shaped by a half-round stamping form. In order to regain their hardness, a large quantity

1. Rolling the steel strips.
Leftover steel
2. Cutting
3. Perforation (first step)
4. Perforation (second step)
5. Tempering
6. Stamping
7. Shaping
8. Hardening (first step)
9. Hardening (second step)
10. Smoothing
11. Grinding
12. Splitting
13. Coloring
14. Finished nib selected as faultless

Production steps in making a Soennecken pen point (1890).

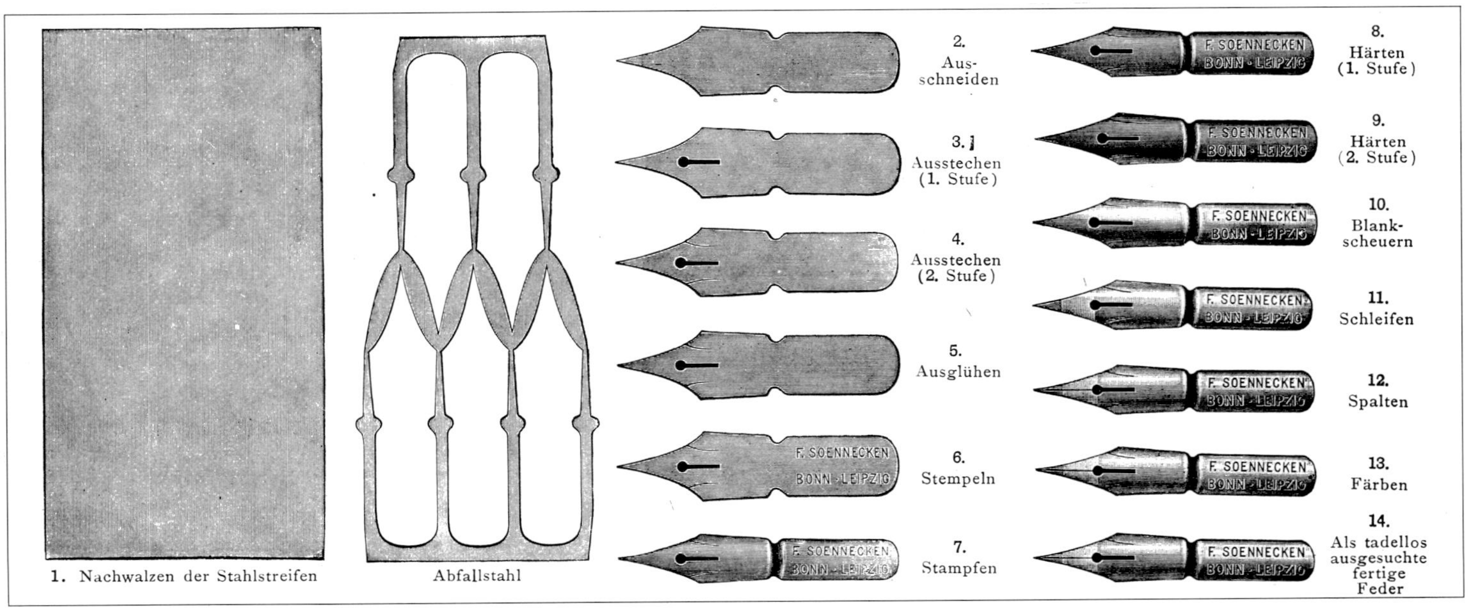

of pens are heated to white heat in flat cast iron pans in the absence of air, and suddenly dumped into deep barrels of oil. Thus the pens take on such a hardness that they spring like glass if one touches them gently. The fat that adheres to them is cleaned off by swirling them in a cylinder that is filled with sawdust. The pens are now heated over a slow fire, whereby the too-great hardness is somewhat tempered for further work. After being cleaned again of the oxides resulting from the heating and hardening by a process of being wiped from both sides for several hours, they are smoothed.

"The smoothing of the pens, which now appear steel-white, on their convex side has the same purpose as the smoothing of quill pens on the back of the pen above the point. A certain part of the metal is made somewhat thinner and more flexible, and the point in which the split is to be made thus becomes elastic and soft. The long slit reaches from the point upward to the vicinity of the hole, while the lateral slit must not touch the point. Then the pens are turned to the right color. This is done by shaking the pens in a turning iron cylinder over a wood fire. According to the degree of heat, the pens first turn a light brown, then dark brown, and finally take on a blue color, and through this process a moderation of the original brittleness is achieved. Only now is the pen split. The pen is placed in a groove so that just half of the nib is firmly in place. The upper stamp of the perforating machine now presses, starting at the point of the pen, on the free-lying half of the

Rolling the steel

◁ *View of the cutting room*

Cutting the sheet metal

Bending the steel pens

Splitting the steel pens

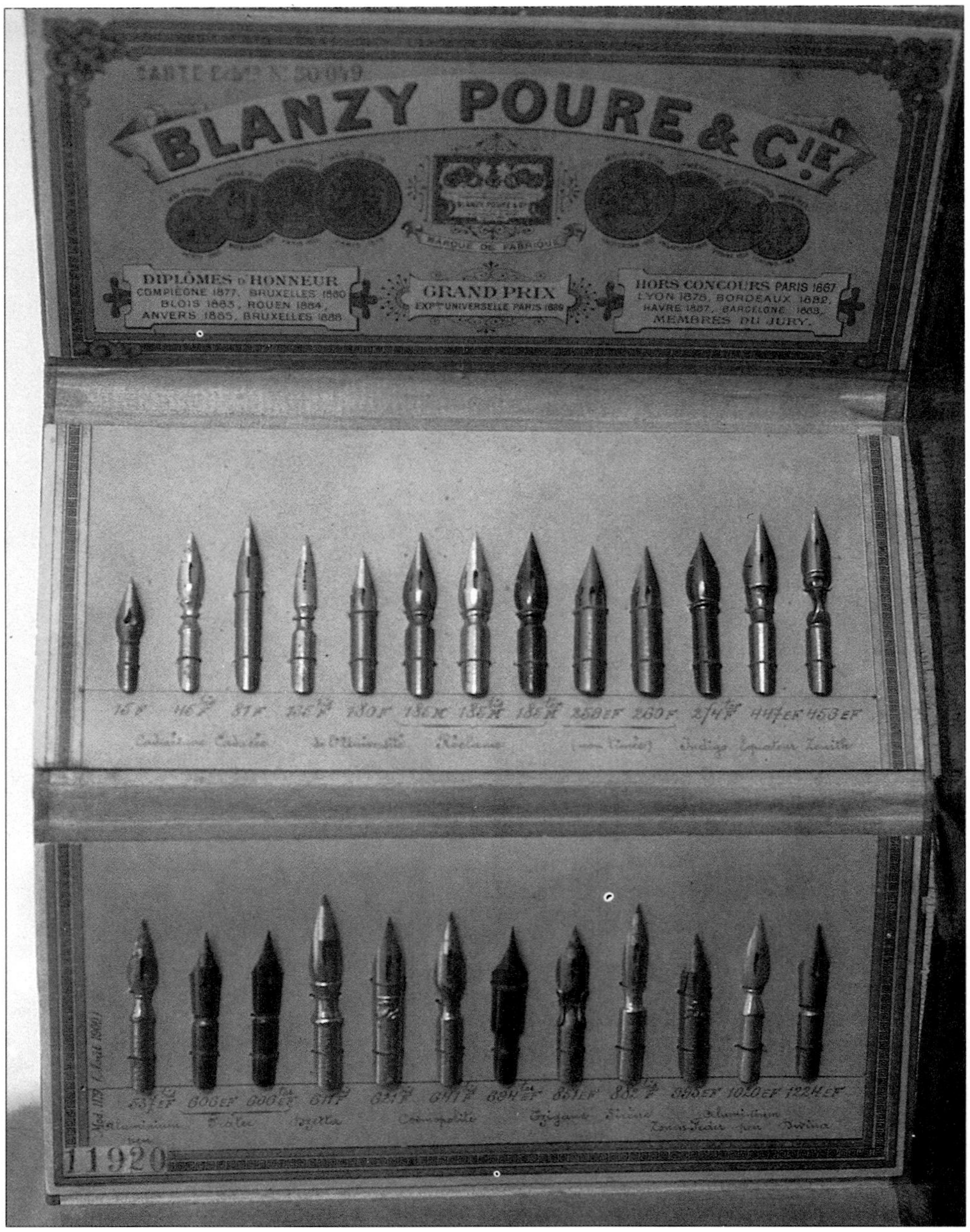

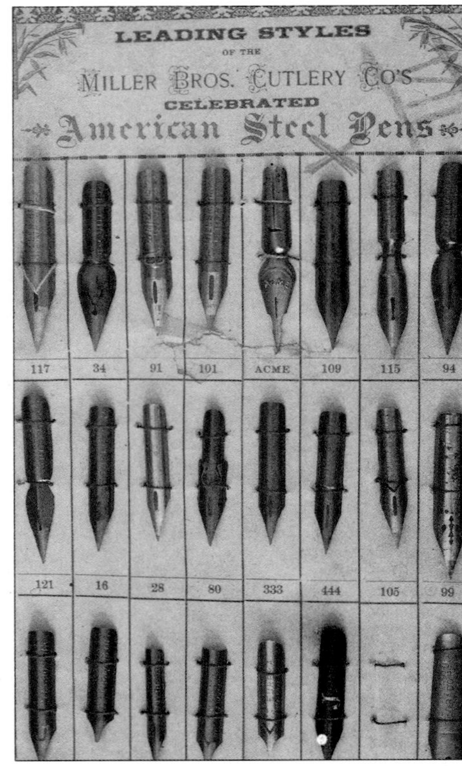

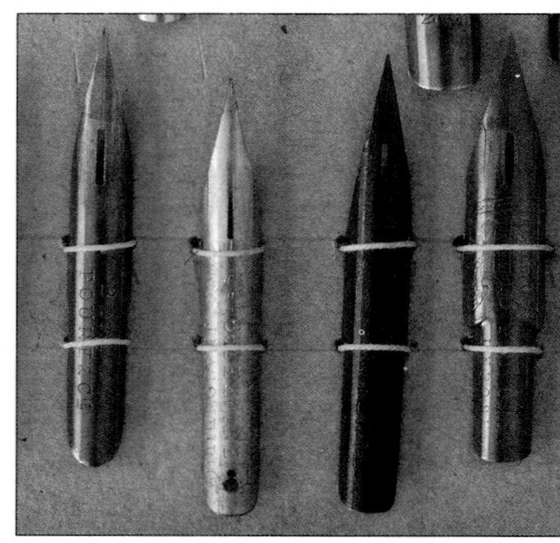

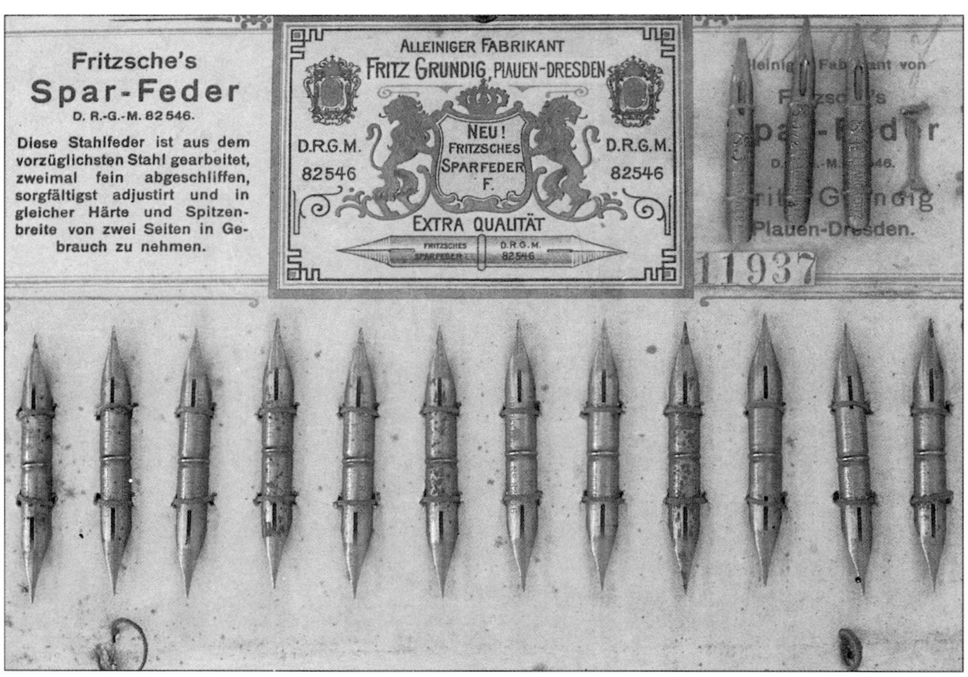

Many wholesalers carried up to 400 different types of pens.

Nineteenth century steel pens from the German Museum's large collection.

pen, downward from the other half, like a shear, and the elasticity is so great that the bent-down, now split part springs back into its previous position by itself. The testing and sorting that now follows demands the best and most experienced workers. Every pen is pressed with its point against a piece of ivory to test it. Those without faults now go to be polished and finally smoothed. Depending on whether the pen is to be sold as a gold-tip, copper, silver, zinc alloy (amalgam), gutta-percha or cement pen in the trade, it is treated galvanically or etched with acid (potassium cyanide) or lacquered."

The success of English steel-pen production spread. In Birmingham alone, there were thirteen pen factories toward the end of the nineteenth century, putting 300 million pens on the market every year. Their success quickly spread to the Continent. In Germany, the Berlin mechanic Heinrich Siegmund Blanckertz developed a power rolling mill in 1852, for the production of fine steel bands to be pressed into pens. Bonn and Leipzig in Germany, and Boulogne-sur-Mer, Aigle, and Paris in France, became centers of steel pen production. Heintze & Blanckertz, Soennecken and Brause were the best known German manufacturers, brands that are, in part, still known today. Large stationers like Roeder in Berlin had steel pens made in England and sold them under their own name in Germany.

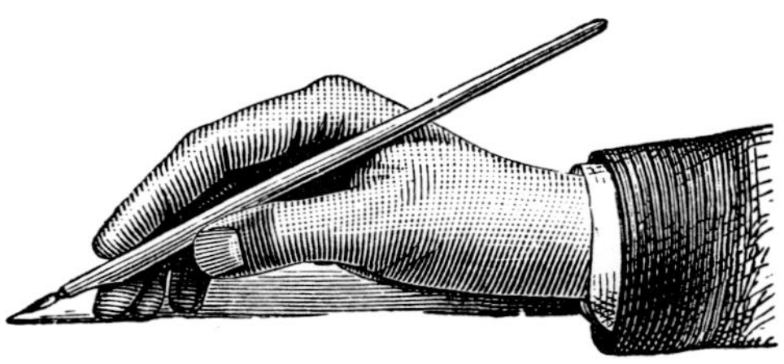

Flat writing position

Along with the many ordinary pens, the number of special pens for various writing needs grew to be incalculable. The most renowned producer of writing materials, F. Soennecken of Bonn, referred in the 1906 edition of his catalog to the "inhuman" variety of pens and offered the following advice for the selection of a suitable pen:

"Is my previous pen well suited to my hand? That is what everyone should ask himself who is concerned about maintaining the writing capability of his hand. We know how hard it is to find a pen that suits the hand. What is the reason? There are too many kinds, too many unusable forms and too many pens in too limited quantities. Among the good and less good steel pens that have been put on the market unsystematically from the beginning of steel pen manufacturing until today in many thousands of different forms, a rational choice is impossible because of the confusion. There is also a general lack of definite principles for the choice of a pen. A mere division into hard and soft pens is not sufficient, and the "new types" that come onto the market yearly with noble sounding names make the confusion all the greater. If this situation is to be improved, then steel pen manufacturers must stop treating man's most important tool as a fashion article and must take the writing habits of the hand into consideration, helping the buyer to make a choice by offeringpractical advice, instead of confusing him with an ungodly mass of types with pointless forms and names." Brightly printed sampler boxes of fifteen pens and one pen stand were recommended as trial purchases: "One will find a pen that suits one's hand among them with no trouble . . ."

Steep writing position

With the S-shaped "normal pen" patented then and the offset penholder that held it, there was a special situation. "The normal pen gives the hand an even position with its diagonal setting, thus making possible such smooth writing as is impossible with the ordinary pen position..., and makes it possible to sit straight in front of the table . . . None of the usual pens will bring out writing so easily and smoothly."

Speed and beauty were also important in the good old days. Fast-writing pens ("glide lightly, never spatter") promised a higher writing speed than the quick pens ("beautify the script and never scratch"). The robust school pen, "the best pen that was ever made for the school," was obviously less hard than the lasting pen for copying purposes, without the typical slit and thus without elasticity.

Notating pens, reckoning pens for bookkeeping that drew two or even three lines of varying strengths simultaneously, round-script pens, and twin pens for decorative writing all belonged to the standard assortment. There were chancellery

Soennecken instructions for choosing the right steel pen: Five basic types with three widths.

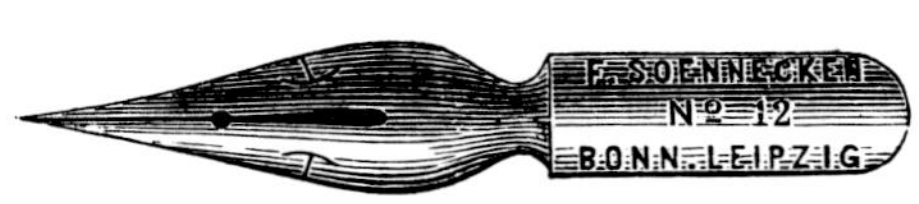

pens, Bremen stock market pens with ball points, ladies' pens, thrifty pens (with an arched cross-section, in order to take up more ink), officials' pens (a combination of school and correspondence pens with a large groove in the middle for holding a lot of ink), the G-pen, the Alfred pen, the Classical (scholar's) pen, the Rasner pen (named after a Viennese writing teacher), the light aluminum pen (aluminum bronze), and not least, the reservoir pen invented in 1878, a direct ancestor of the fountain pen. The underside, pocket, and beak of the pen held the ink supply for a few extra lines before it had to be dipped into the inkwell again.

In the German schools three types of pen established themselves toward the end of the nineteenth century: the Ly-pen, angled to the left, the To-pen, angled to the right, and as of 1911 the ballpoint pen (platter pen), Ato- and Redis-pens. All later writing and drawing pens were based on these three basic types.

Over 400 different types of pens were stocked by most wholesalers, in the reigning colors of the season—silver, copper, black, bluish. And whoever was ready to pay ten to twenty times the price of a customary steel pen could give his thoughts and effusions their proper value with an elegant silver-plated gold pen with a "diamond tip." In this case the diamond was nothing more than a soldered-on splinter of osmium-iridium, a discovery that is ascribed to the Frenchman, Mallet, in 1843. The "diamond tip" gave the pen the desired durability, for the platinum metals osmium and iridium, which are welded onto the better writing pens to this day and then finely polished, practically never wore out. Gold pens consist of a very hard, elastic alloy of, at most, 14 parts of pure gold plus portions of silver and copper. The 14-karat pen predominates, but 18-karat pens (especially soft) and 12-karat pens (hard) are also available.

The simple steel pen did not remain untouched by the apparently unquenchable human longing to beautify all its possessions. The pen manufacturers competed, using originality and imagination, to create the most decorative fashion pens. The hole required to hold the ink was changed into slits, crosses, triangles, anchors, half-circles. Additional stampings formed faces, hands with outstretched index fingers forming the tip. Souvenir pens brought back memories of special events or military victories. Portrait pens portrayed the faces of important personalities from politics (the Bismarck pen) and art, or showed emperors and kings. Devotional pens with crucifixes kept alive the memory of successful visits to places of pilgrimage. Whoever wanted something earthier might be contented with flowers, animals or even ponderous trains and ships on the steel pens that glided lightly over the paper. These ornamental pens, complete with fine pierced patterns, seem quite superfluous to us today. They were completely unsuitable for writing. One gave them as presents, for remembrance.

Though at the beginning of the age of steel pens it was still sufficient to stick the writing nib into a cut-off quill, the demand for devices that were more comfortable to hold soon led to the development of the penholder. The penholders of the nineteenth century were made of finely polished walnut, Tonkin reed, cork, rush, horn, even artistically carved mother-of-pearl and ivory. In the realm of the metal penholder, things became more elegant still, and more expensive. Often penholders were silvered or gilded, with a button-like cap or a loop for fastening to a watch chain or necklace. Many penholders could be made longer for use and then telescoped together to carry comfortably in a jacket pocket. The ultimate may be a design that had an extending tube with a fine pencil lead under the pen. So the combination writer is by no means an invention of our own time.

Then as now, the taste of the general public inclined toward elegant, thin penholders. In everyday use, especially for people who wrote a lot, such as chancellery and office employees, many a slim penholder proved to be an instrument of torture. The hand cramped painfully, and the required fine artistic script, such as one sees in copperplate examples, was no longer attainable.

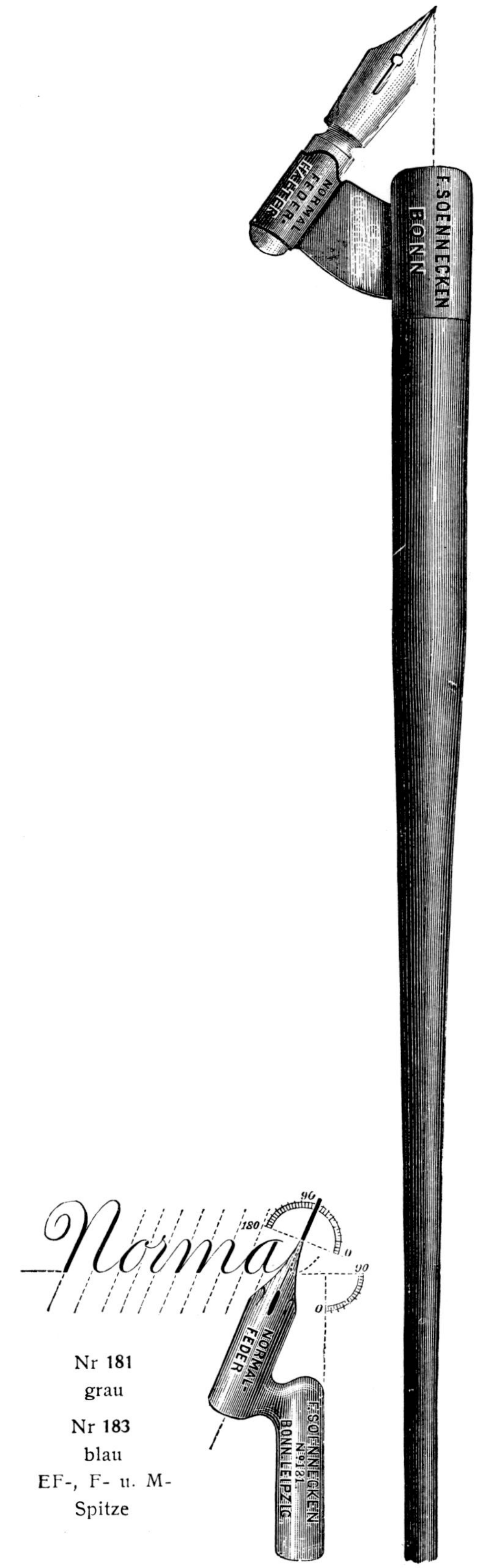

The "normal pen" allowed the writer to sit straight in front of the table.

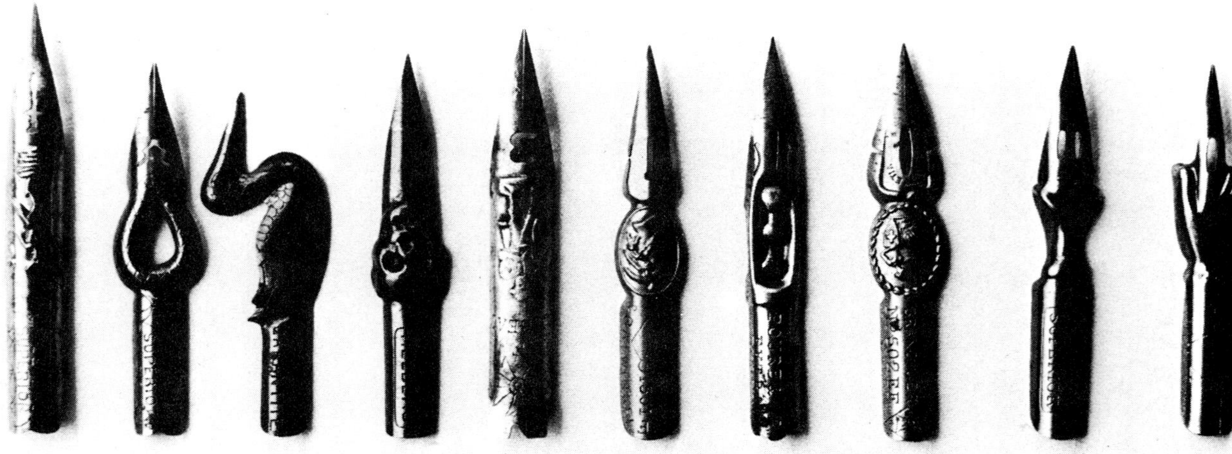

A variety of steel pens, some with rare forms.

Cut penholders made in Brienz around the turn of the century.

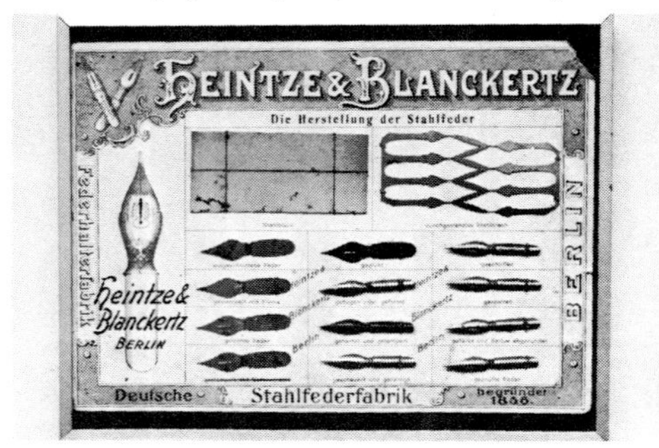

Colorfully printed cardboard boxes held the desirable steel pens.

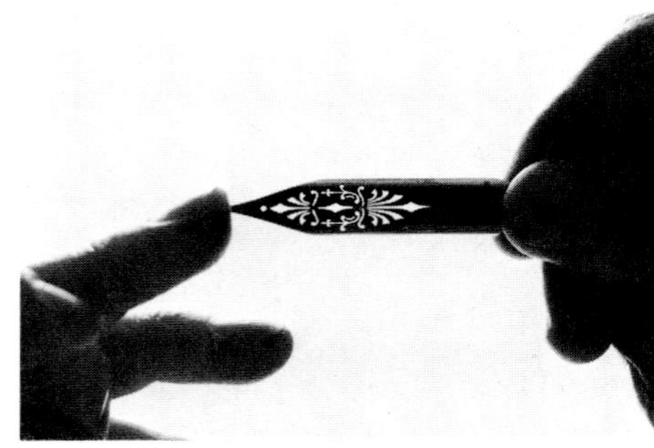

Ornamental pens, intended not for writing but as souvenirs.

Turn-on-the-century penholders.

Soennecken's gold penholder with telescopic pencil under the nib.

"Whoever wants to protect himself from writer's cramp, or feels the first indications, a cramp-like weariness of the fingers while writing, should choose a light, thick penholder without metal . . . and get used to using the quick-writing pens that write without the application of pressure . . ." recommended the pen manufacturer and dealer F. Soennecken around the turn of the century, while simultaneously offering concrete objects to make writing easier. These included penholders with cucumber-shaped thickenings, whose points could be adjusted in various directions, a ball of cork the size of a tennis ball with an inserted penholder, or a "penholder grip of cork in the shape of an egg, which has a weight of 250 grams thanks to a metal insert and offers all the fingers a full hold . . ." The ergonomic ultimate seems to have been the penholder form that filled the hand completely and was equipped with grooves for the individual fingers.

These efforts to avoid the unpleasantness of writer's cramp knew no limits. It is only surprising that none of these imaginative devices managed to catch on and last. Obviously, the less voluminous but better-designed and tested handles of present-day school pens are better suited to avoiding writer's cramp. Regarding school penholders, the fight against a cramped writing position and body position among German students was also supported by efforts to provide technical help in the best sense of the word. Soennecken's writing aids were praised by students and doctors, parents and teachers as "simple and comfortable means of preventing the improper body position of the pupils and the resulting bodily weaknesses, such as nearsightedness and curvature of the spine." The "before-and-after" examples must have enlightened the concerned pedagogues and parents as well as the scientific community, for unsolicited testimonials from royal government offices at Königsberg, Merseburg, Frankfurt, Würzburg, as well as from many well-known eye doctors and school administrators were received.

Soennecken's penholder designed to eliminate writer's cramp.

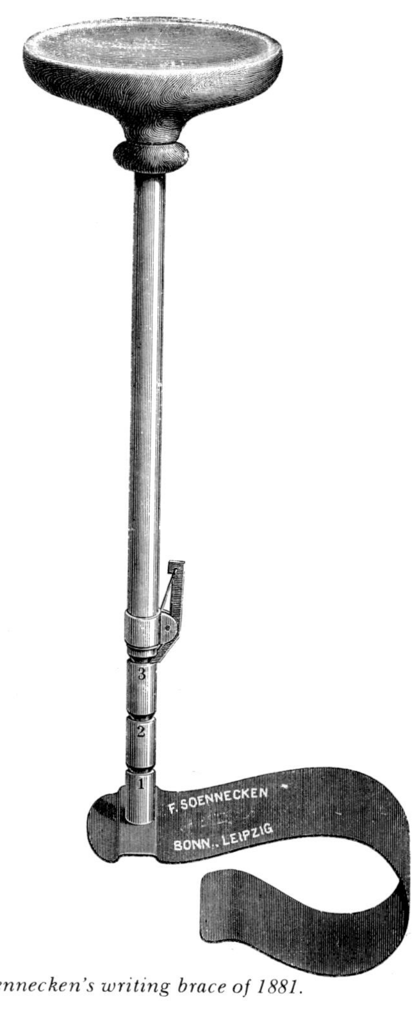
Soennecken's writing brace of 1881.

Before

After

Picture Postcards

The poor man's letter, the postcard, was introduced in Germany only in 1870. It was a complete success from the start. Because the postcard offered only little space for writing, the writer could limit himself to a few lines. And what was even more important, the expenditures for expensive writing paper, envelopes and perhaps even a wax seal on the back, were spared.

The desire to share some sort of illustration as well as a personal message was completely fulfilled by the idea of the picture postcard. Later other types, such as pictures of groups in peasant costume and comic cards, made their appearance. The imagination of the publishers around 1900 was almost unbounded. It is no wonder that the world began to collect picture postcards—used or brand new. The collecting hobby began with the middle-class people whose well-situated members and friends could afford to travel and were already sending home postcards that evoked prestige, not to mention jealousy. The socially less well-off classes soon followed the example of the more affluent. Young soldiers on duty far from home were just as zealous writers of postcards, picture and otherwise, as journeymen in training.

The outbreak of World War I promoted the use of postcards in military mail. Naturally there could not be picture postcards of battlefields—pictures of which would scarcely have served to encourage the desired general enthusiasm about the war. Less military but sufficiently patriotic motifs were printed on postcards that were distributed free to the troops fighting far away from home.

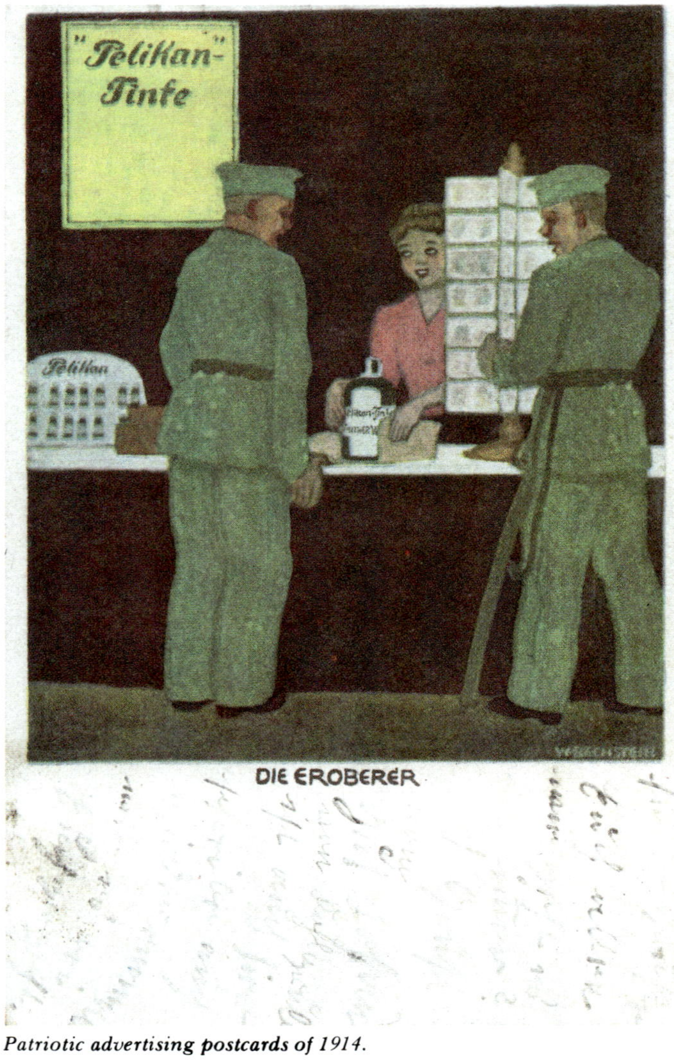

Patriotic advertising postcards of 1914.

Handwriting

Graphology originated in the middle of the nineteenth century. In his groundbreaking studies, the French abbot Jacques-Hippolyte Michon referred to information that every person's handwriting contains for the first time.

The most noteworthy German exponent of graphology was the philosopher Ludwig Klages. It owes its scientific basis to him. Graphology is studied today at universities, medical and psychological trainin. he accuracy of modern graphology is about 90%. The trained graphologist views handwriting "as an entity." He analyzes its movements, combining the results with psychological knowledge. Handwriting shows him something of the writer's emotional life. A swinging, unhampered script indicates a strong, happy personality. It is just the reverse in terms of sadness. The person is "as if crippled," a "pile of misfortune," on the decline. Correspondingly apathetic lines appear in the handwriting. Anger lets inner tension show in exaggerated expression, and the pen seems to have charged across the paper.

Handwriting analysis is often made use of in processing job applications. Often a sample of handwriting, usually the resume, is requested, a clear sign that in cases of doubt an analysis will be done. There are three areas that particularly interest the personnel director: potential for accomplishment (strength of will, creativity, talents), ability to fit in (sociability, team spirit), and trustworthiness (loyalty, respectability). Of course one makes sure not to base one's judgment of the applicant on the handwriting analysis alone. A general rule is: 15% references, 15% documents, 20% "grapho" and 50% personal interview.

Now and then a lefthander who was forced to write right-handed as a child will break the rules of graphology. Experienced graphologists then depart from their usual methods and ask the applicant to copy a few lines out of a newspaper, instead of the resume, to which one usually devotes particular care and writes neatly and legibly.

Naturally handwriting changes over the course of time. A handwriting analysis cannot be valid forever, for the personality, and therefore the handwriting, will undergo changes.

What writing instrument should one use for handwriting that will be judged? The most unsuitable types are the pencil and felt-tip pen. The uniformly heavy script gives too little evidence for a responsible evaluation. The once-scorned ballpoint pen is not much more highly regarded by the graphologist. The pressure of the writing can indicate dimensional depths, and specimens of writing done with a fountain pen give the clearest evidence. And no writing instrument expresses as many nuances as the flexible steel or gold pen.

The handwriting lets the graphologist see into the personality.

At right:
Weak-willed, inept married man
Strong-minded, spiritually unclear housewife
Self-assured man

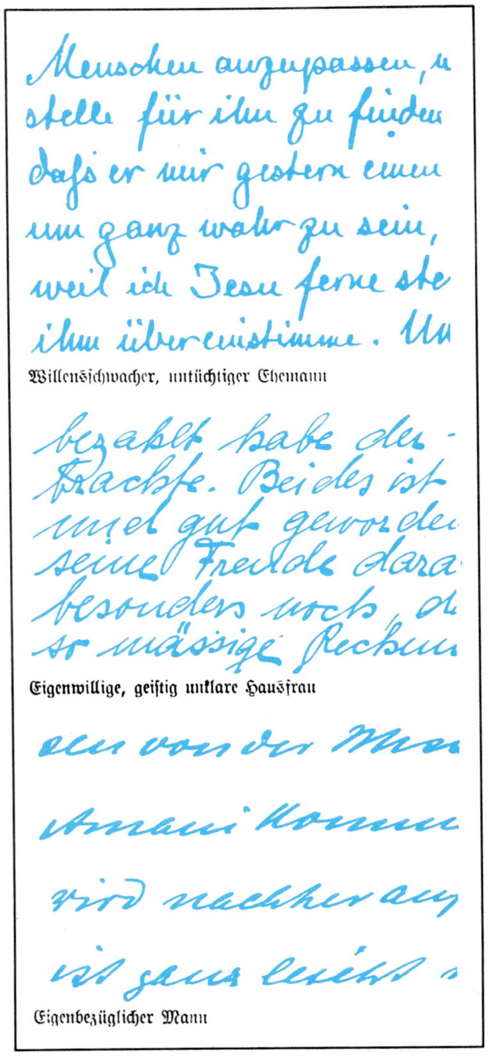

Graphological texts

No other writing instrument allows a clearer judgment of a person's writing than the fountain pen.

F. Soennecken

Soennecken's gold fountain pen with diamond (iridium) nib of 1906, "to be carried in any position you wish."

The founder of one of the oldest German writing instrument companies, Friedrich Soennecken, was not only a successful businessman but also won lasting fame as a proponent of handwriting. He was born in Dröschede, near Iserlohn, in 1848, the son of a blacksmith. He attracted the attention of his teacher, the enthusiastic calligrapher Rahlenbeck, as a schoolboy. His schoolmaster's influence was to be a major factor in the young Soennecken's later professional success. After commercial training, Soennecken, eager to learn more, founded the Merchants' Society for the Promulgation of Knowledge. It was particularly the fine handwriting taught by his teacher that inspired him. He gave lectures on decorative handwriting and tried to make further developments in the round script that had originated in France. To promote this simplified new writing system, his firm, F. Soennecken's Publishers, which he founded in Remscheid in 1875, put out small brochures, which had already reached their hundredth issue by 1879. Dissatisfied with the steel pens available at the time, he had special round-script pens made in England according to his own ideas.

In 1876 he moved his young firm to Bonn. The university atmosphere excited the man who was eager to learn. Here he could continue his studies. Along with his activities as a publicist, he made many inventions that made office work easier: the traveling copier press (1881), special ink bottles and the 111 school pen (1884), letter organizers, paper punches, and office furniture (1886). Along with his innovations, splendid in the best sense of the word, is the first German penholder with a gold pen (1890).

Soon F. Soennecken ranked among the most important producers of steel pens, penholders, fountain pens, ink bottles, letter organizers, office furniture, copying presses, erasers, moisteners, briefcases, letter racks, calendars, compasses, and drawing and typing tables—all in all, more than 5000 different articles used in writing and office work.

Soennecken also made a name for himself as the champion of a simple assortment of writing pens. The writing-pen manufacturers of the turn of the century, he found, simply offered too many different types of pens, so that no one could select a type of pen suited to him out of the immeasurable vastness. The consumer was led to make his choice not by the suitability of the pen, but by aspects of fashion. So Soennecken limited his offerings to five basic types of pens with three point widths. A pen positioner invented by him guaranteed the user the right, comfortable writing angle at all times.

The first gold penholders with "diamond-iridium points" were generally telescopic pocket pens with an integrated pencil under the pen, and the fountain pens were simple devices made of hard rubber, with a tongue either over or under the point. In 1905 there appeared the "safety fountain pen—that can be carried in any position," with a short cap that screwed on tightly. In 1924 there came the first lever fountain pen, *Utilitas.* About seven years later, Soennecken's assortment of fountain pens in the modern style made their appearance. Along with the immortal safety pens, now pressure pens such as the famous *Rheingold* were offered. In 1938 the *President* and *Rheingold* pressure fillers, both with ink-level windows, were joined by the first Soennecken piston filler.

Meanwhile the Soennecken firm had passed to the second generation. After Friedrich's death in 1919, his son, Dr. Alfred Soennecken, had taken over the direction of the world renowned firm. Under his leadership the number of employees doubled from 1034 in 1914 to almost 2000 (!) in 1938. With much difficulty, the Soennecken factory, badly damaged in the war, was rebuilt in 1945: The Soennecken Piston Fountain Pen III,so-called in memory of the famous school pens of 1884, was available in three sizes and seven colors and ranked among the best-sellers of the postwar era.

SOENNECKEN'S
KURRENTSCHRIFT-
FEDERN
No. 106
F. SOENNECKEN'S
KURRENTSCHRIFT-F.
BONN · LEIPZIG
GESETZL. GESCHÜTZT
FABRIK-MARKE
GESETZL. GESCHÜTZT
FABRIK-MARKE
GESETZL. GESCHÜTZT

In 1967, after 92 years of existence, the ambitious firm suffered financial difficulties and had to be liquidated. The remains of the business, consisting mainly of the trademark and patent rights, became the basis of a new firm founded by the Electric Light and Power Facilities Company in 1979: the Soennecken Office Organization GmbH. This firm was taken over in 1983 by the Wholesale Union of German Office Supply Companies eG. This firm has since brought the Soennecken brand back to new life.

Today a variety of office articles, bearing the modernized Soennecken trademark, is available to the Federal Republic of Germany and, under license, in many countries around the world.

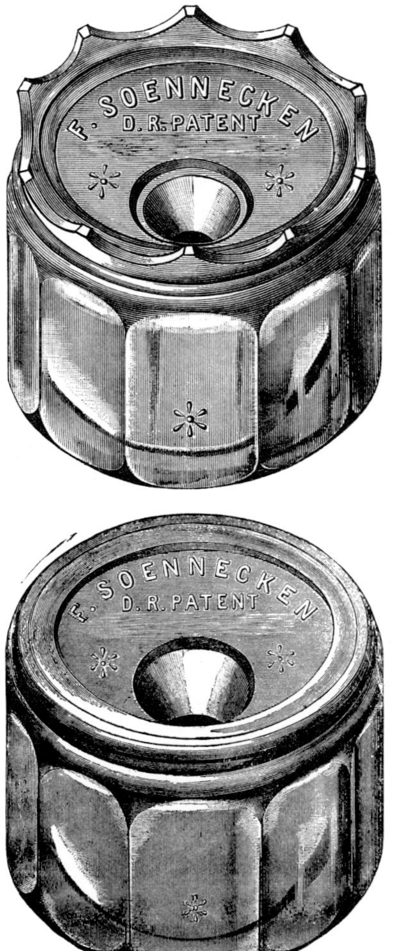

Soennecken's inkwells "prevent too deep dipping of the pen," 1890.

A 1952 advertisement.

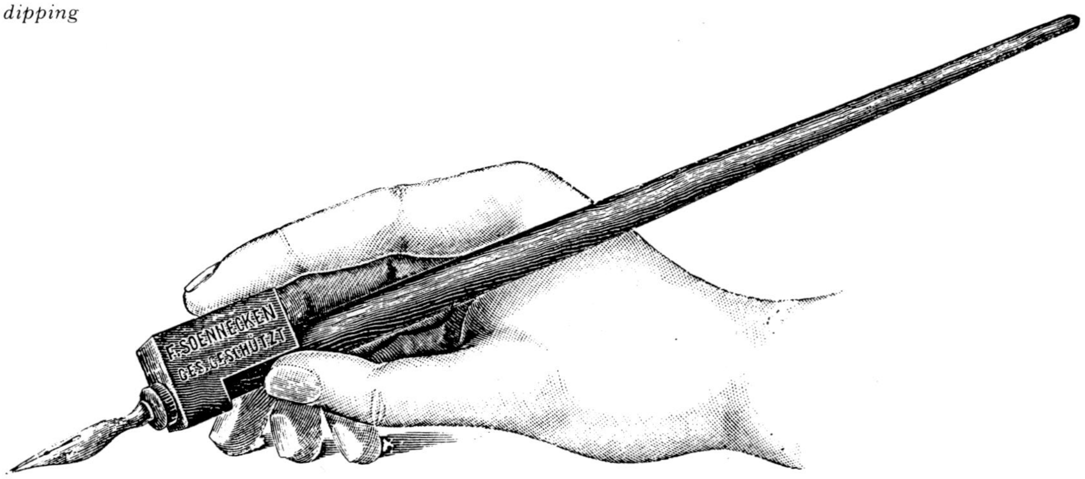

Soennecken's penholder with cork finger rest.

Color picture, page 43:
Current-(fast-) script pens were band pens. With them, "even an otherwise irregular, unpleasant handwriting" gained "an astonishing regularity and beauty, which the writer cannot produce with any other pen." From Soennecken's 1890 catalog.

Christmas catalog of writing instruments, 1932

F.S.
Soennecken
schenken —
heißt zeitgemäß
schenken!
SOENNECKEN-Füllhalter
Höchste Zweckmäßigkeit, vollendete Eleganz, echtes Material – das sind die Vorzüge, die den SOENNECKEN-Füllhalter als Weihnachtsgeschenk besonders geeignet machen. Viele Größen und Ausführungen – auch mit Edelmetallumkleidung – in verschiedenen Preislagen. Ausführliche Prospekte kostenfrei
SOENNECKEN-Füllflaschen
Aus Hartgummi, fein poliert
Praktisch, standsichere Form
Nr 901 8×5 cm ... RM 3.—
Nr 902 9×6 cm ... RM 3.50
Nr 9: RM 12.50
SOENNECKEN
Nr 810: RM 14.50
Nr 814: RM 19.—
Etwas kleiner = Nr 813: RM 17.—
Mit Gold-Eilfeder Nr 820 E: RM 25.—
Selbstfüller Nr 710: RM 14.50
SOENNECKEN-Garnitur „Rheingold"
Eine besonders schöne und elegante Geschenkpackung, bestehend aus einem Druckfüller Nr 916 und einem Drehstift Nr 140. Innenausstattung des Etuis: Atlas und Samt
Garnitur Nr 1614: RM 28.—
SOENNECKEN-Drehstifte
Schwarz und farbig
Nr 128: RM 2.75
achtkantig, 12 cm lang
Nr 140; 14 cm, rund RM 2.50
Druckfüller „Rheingold"
Nr 916: RM 25.—
D·R·P ang.
DRUCKFÜLLER „RHEINGOLD"
Ein Edelerzeugnis deutscher Füllhalter-Industrie. Technisch vollendetes Druck-Füllsystem. Ein Fingerdruck genügt, um den Halter aus jedem Tintenfaß zuverlässig zu füllen. Schwarz und in wundervollen Farben
F. SOENNECKEN ★ BONN
BERLIN W MOHRENSTR. 58/59 ✶ FERNRUF: MERKUR 1086/1087 UND 5342/5344 / LEIPZIG MARKT 1 (ALTES RATHAUS) ✶ FERNRUF: 14239
SOENNECKEN-ERZEUGNISSE sind in allen Schreibwarenhandlungen erhältlich

The Pencil

Dürer's silver-pen drawing: "Self-Portrait of the Thirteen-year-old Albrecht Dürer," made in 1484.

The down-to-earth writing and drawing instrument for everybody, the eternally young pencil, certainly does not belong among the noble writing tools but is the most frequently used of them. Unlike most writing instruments, it does not put any colored fluids on the paper, but rubs itself off for the user. Its closest relatives and ancestors were very popular even in the earliest days of mankind. Pieces of chalk, madder and coal were the ancestors of the lead-gray and other colored pencils.

A pencil of actual metallic lead was already used by the Romans to write on wax tablets and to draw lines of papyrus and parchment. Whoever could afford to also used "steffte" of lead, tin, or silver in the Middle Ages. The disadvantages of being pale in color and easily erased were accepted in exchange for the advantage of being ready to write on paper. The Nürnberg artist Albrecht Dürer made the silver pencil famous. Among his more than 1000 drawings, the "self-portrait" of the thirteen-year-old boy stands out as the most noteworthy silver pencil drawing. But Van Eyck, Botticelli and Leonardo da Vinci also created great works with the thin bar of a lead or tin alloy.

The difficulty of working with a silver pencil compelled a search for a better blackening material. Around 1400, mountain farmers in the Bavarian Forest found a hitherto unknown, very black material but did not recognize its value as a writing material. A century later, in 1564, chance helped out. English shepherds in the hills of Cumberland, near Borrowdale, found blobs of blackish metal under torn-out tree roots after a thunderstorm. The practical shepherds thought of a use at once. The material was ideally suited for marking their herds of sheep. At first the discovery was thought to be lead ore, and was used to make casting forms for cannonballs. Later the substance, called "black lead-white" because of its shiny silver surface, was cut into thin rods, pointed, inserted in wooden or metal holders and sold as a writing instrument. Thus did the English pencil industry, which was soon to attain a worldwide monopoly, have its origin. Only in 1789 could the Swedish chemist Carl Wilhelm Scheele prove that this "lead-white" was composed of the same element as its equally pure crystalline counterpart, the diamond. Now one might think it would have been inevitable to rename the writing material "graphite pencil" (Greek "graphos" = to write), as the mineralogist A. G. Werner suggested in 1799, but the term "lead pencil" (German "Bleistift") was so deeply rooted in general usage that, even in the face of better knowledge, it has not been possible to displace it to this day.

The non-toxic graphite, that made such a black line, was soon mined from the state-owned Cumberland Mine to the point of exploitation. On pack horses and in wagons, accompanied by heavily armed escorts, the unique, valuable material (one kilogram was worth 1000 English pounds) made its way from Borrowdale to the arsenal in the Tower of London. This tempted English miners to smuggle small quantities out and take them over the hills to the coast on dark nights, selling them to Flemish dealers who transported them to the Continent, thus breaking the English graphite monopoly.

At the end of the eighteenth century, the deposits ran out, despite limiting mining to only six weeks a year. The English Parliament forbade the exportation of graphite, supposedly on pain of death. Gradually the quality of the English material grew worse, as did that of the pencils made of it. To stretch the graphite, binding materials such as glue, gum, tragacanth, etc., were officially added. Because competition on the world market was still lacking, England could still export these poorer pencils at high prices.

It was not long before mines in Passau, Bavaria, and in Bohemia were also producing the desirable "lead-white." Because of the crude mixtures of quartz, feldspar, and other minerals, it could come nowhere near equaling the purity of the

A lead-white cutter in a 1711 instruction book. Graphite was then regarded as a shining white lead metal—hence the term "lead-white pencil-maker."

"It is a real pleasure when the nice lead glides over the white paper, from gleaming black to the finest silver tones, and interrupts neither the writer's nor the reader's thoughts, but moves them along on a gazelle's hooves."

from "The Garden House," 1861

English graphite. Inventive as the Nürnbergers were, they looked for a solution, and they were able to develop a method of producing "artificial" material. The pieces of graphite produced in Germany were pulverized in mortars and rid of their coarse impurities. The resulting "graphite meal" now had to be bound together again and made capable of writing. The "lead-white smelters" added molten sulfur, gum arabic, or resin. They then formed flat cakes and cut rods out with a saw. The art of producing artificial pencil leads remained a secret for a long time. Only in 1772 was the method made public in the book "Handicrafts and Arts:"

"He pulverizes the fine watered lead in a mortar, and since this always contains foreign mineral pieces such as sand, he sifts it two or three times, every time in a finer hair sieve. In a crucible there is then added, for the time being, one-quarter pound sulfur for every pound of watered lead. For the German watered lead is not usable in its usual state. As soon as the sulfur is liquid, the pencil maker pours in the watered lead in the indicated proportion, and the sulfur encourages the melting of the watered lead. But he must be careful and see that the flame does not reach into the crucible and set the sulfur afire. For that reason the mixture is often stirred. When the watered lead is evenly liquid, it remains in the crucible with its mixture, with which it unites, until it has cooled to the point at which it is still liquid but can be touched with the fingers without injury. In this condition it is poured out of the crucible onto a table or a board and pressed flat as a pancake with the fingers. The pancake must cool thoroughly before it can be worked further. This working is very simple. The pencil maker cuts the pancake with a fine-toothed saw into small plates, and out of these he cuts the pencils with the same tool, to their established size. Each pencil may now be set in wood."

The graphite bars so produced wrote worse than they should have. A few black sheep among the pencil makers thus got the idea of using English graphite at each end of the pencil, but material produced in Germany in the middle. It was customary that the cautious buyer would hold the pencil point over a flame to test whether it was mixed with sulfur. Pure graphite, such as that from England, does not burn.

The artisans of the free city of Nürnberg perfected the art of "lead-white cutting" and insertion into grooved wooden staffs. This does not seem to have proceeded peacefully. In 1662 official action was noted concerning disagreements between representatives of the established cabinetmakers' occupation and the "lead-white pencil makers." The latter practiced their trade in the guild-dominated city as a "free handicraft," a non-organized trade. At first the work was divided: the pencil makers made the wooden casings, the lead-white cutters sawed the graphite blocks into pencil leads. Naturally the cabinetmakers wanted to be part of the new trade and tried to gain control over the whole business. They accomplished, for example, the banning of Friedrich Staedtler from making pencils, according to a city government decree of 1662. He had committed the offense of carrying on by himself what later became mass production. With the help of cutters, he cut the graphite into small rectangular bars. They were then placed loosely in holders, artistically formed by cabinetmakers, and glued in place.

Though the cradle of German pencil production was in Nürnberg, the Nürnberg pencils made of native material were, at least in the beginning, not of typically German high-quality. The casings were made of linden and pine wood, the more expensive cedar wood being used only later. Production, according to a contemporary and accurate description, took place "by gluing grooves and casings together." The "grooves" were originally rectangular pieces of wood, in the middle of which a lengthwise groove had been cut with a so-called "sticking plane." The groove needed to be only deep enough to be completely filled by the lead rod. When

De figuris lapidum, &c.

Stitij puto, quod aliquos Stimmi Anglicum vocare audio) genere, in mucronem derasi, in manubrium ligneum inserto

.2.

.1.

L. Lateres e luto finguntur & coquunt, ad ædificiorum parietes, pauimenta, caminos: item ad furnos, aliosq; vsus.

Graphit, der neue Schreibstoff

In a book published in 1565, graphite (Stimmi Anglicum) was first discussed as a writing substance.

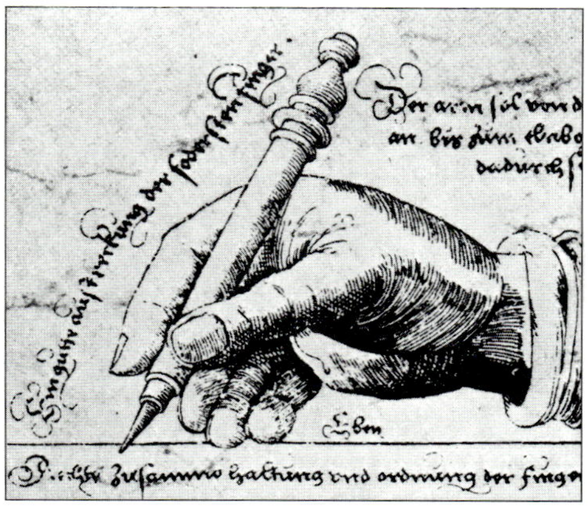

The first pencil leads were set in a wooden casing.

To stand above his competitors, Lothar von Faber created one of the first brand-name pencils, hexagonal and green.

Grinding graphite and clay

Mixing the raw materials

Preparing the "lead"

Sawing and planing the wood

Gluing in the leads

Planing the casings

Polishing and lettering the pencils

Tying and packing

Scenes from a pencil factory. Work processes for men and women were strictly separated.

this was inserted, "a casing was glued to the piece, so that the graphite was now completely enclosed . . . thus we have here a rectangular pencil before us, which takes on its round form only after being worked with the rounding plane."

The pencil owes its essential qualitative improvement to the continental blockade that Napoleon set up against England at the end of the eighteenth century. Cut off from the English graphite mines, the French looked for an alternative and gave an artist and officer in Napoleon's army, Nicholas Jaques Conté, the job of developing a synthetic material. In 1795 he was able to mix graphite with plastic clay instead of sulfur and make pencil rods that attained a hitherto unknown hardness by being heated to a high temperature in the absence of air.

The new method of production brought two improvements of great importance for the future. By changing the proportions of graphite and clay in the mixture, leads of different degrees of hardness could be made for any desired use. The grades of hardness were indicated by the traditional English indicators that are still used: H = hard, B = black = soft, F = firm, HB = half black = medium hard. The greater the proportion of graphite, the blacker and softer the mark! In addition, the new production technique made possible a much more sensible use of graphite and thus built the foundation for a real pencil industry.

At the same time, independently of the French, the Austrian Hardtmuth had invented a similar process, by which he could turn out high-quality pencils. That was urgently needed, too, for the Nürnberg "stumps" had already given the product as a whole a bad reputation. Around 1800, products that only looked like pencils left the Nürnberg workshops. In reality they were solid pieces of wood whose two ends were tipped with graphite to look as if they contained a graphite rod.

Lothar von Faber improved Conté's process. To be able to rise above the cheap "Nürnberg wares," Faber created one of the first brand names in Germany for his quality products, "Faber pencils." The other Nürnberg pencil manufacturers did the same. Names still famous today, such as Staedtler, Schwan, Lyra (Froscheis), and Guttknecht formed the basis of the German pencil industry, whose products have been sold around the world. Weekly production in 1921 added up to:

Manufacturer	Founded	Production
Johann Faber	1878	18,000 gross
A. W. Faber-Castell	1761	14,000 gross
Schwan	1855	9,000 gross
Staedtler	1839	5,000 gross
Kurz	1882	2,000 gross
Lyra	1806	7,000 gross

Today Faber-Castell alone produces more than a billion wooden pencils a year.

Over the years the pencil remained the reasonably priced yet uneconomical writing instrument. Through repeated sharpening, two-thirds of the lead will be wasted during a pencil's life span. It was not long before wooden-cased leads were enclosed in cylinders of metal or horn from which they could be pushed out. In 1869 the American A. T. Cross developed the first all-metal pencil, whose graphite rod was contained in a metal tube and could be moved in and out. Further types that did not need to be sharpened followed.

The mechanical pencil allowed the lead to be pushed out by turning a mechanism at one or the other end of the pencil. This type, in miniature form, was very popular among young ladies as a "dance-card pencil." With it they could write their partners' names on their dance cards.

In the mechanical system (the so-called Eversharp system), the lead was inserted in a turning point. By turning it to the right, the lead was made to come out, but it could not be retracted again. After disengaging the mechanics, the lead could be pushed back into the tube by light pressure.

Traveling vendor with "rare English pencils," circa 1810.

The "fall pencil" opened a mechanical pincer when a button was pressed, letting the lead slide out. This type of pencil is still preferred today by technicians and artists.

Its perfection for the time being came only with the one-hand pressure pencil that appeared at the beginning of the twentieth century. It had enough storage space to hold up to nine leads, which were moved to the point one after another automatically at the push of a button. The amount of lead available was enough to draw a line one-and-a-half kilometers long.

There were also, and still are, shaking pencils that, when moved backward, let the lead move forward. The little protective filler pencils were neat: instead of a clip, they had a tassel of colorful silk cord. This is how a pressure pencil of the 1950s functioned: "At the push of a button, the conically formed head of the moving mechanism is pushed out of the holder. When it emerges, the tongues of the moving mechanism, formerly pressed together, spread and free the lead that has been pushed forward. The front pincer, which likewise moves the lead one writing length forward after being pushed, brings the lead out at the point and prevents the lead from falling out during the process. When the push-button is released, the formerly tense spring pushes the mechanism back. In the process, the moving mechanism moves back into its stopped position in the casing. The tongues of the moving mechanism are pressed together, holding the lead for the next push forward."

Along with pencils for lead of a single color, some firms also offered filler pencils with several colors. With the help of a lever on the side, any one of four or six different-colored leads could be made ready for writing.

Again and again, engineers tried to simplify the operation of the mechanical pencil. In 1956 Pelikan introduced the "bending pencil." Using a push-button would interrupt the flow of writing, it was said. It was simpler and more convenient, though, to bring out the next piece of lead without changing one's writing position—simply by pressing one's thumb. In the process, the Pelikan 60 bent neatly in the middle.

The fine-lead pencil of today has scarcely anything in common with the comparatively clumsy filler pencil of past decades. In place of clay as the binding material for the coloring matter of the lead, various synthetic materials (polymers) are used, depending on the degree of hardness. "High-polymer leads" are produced for line widths of up to 0.3 mm, and are flexible and especially unbreakable on account of the extraordinarily high concentration of their molecules. The mechanics of the present-day fine-writing pencil allow a variety of writing comfort, depending on the use: "Fixed pipe" (stiff-tube) types for drawing with a ruler or stencil; "half slide" (half-extending lead tube), a combination of the drawing and writing pencil, "full slide" (fully lowering tube) for writing. Here the tube comes out of the housing at the press of a button and brings the lead with it. Since the lead protrudes only a small fraction of a millimeter, it is protected against breaking off. During use, the tube automatically slides back into the housing, corresponding to the lead consumption.

For the time being, the high point in the development of the mechanical pencil has probably been achieved by A. W. Faber with the TK-Matic: The world's first non-stop fine-lead pencil, it writes or draws 180 meters without pressure. After one lead is used up, the next one moves out of the reserve chamber into the automatic mechanism.

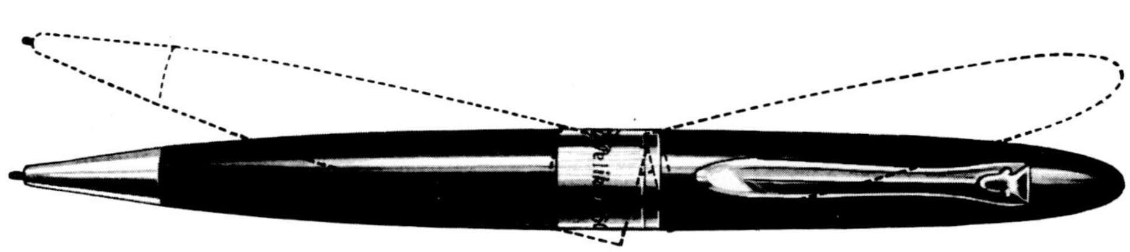

The Pelikan 60 jointed pen. This pressure pen, made since 1956, bent in the middle to carry a lead.

The Pencil in Literature

The "eternally young pencil," this unlikely writing and drawing instrument for everyone, surely does not rank among the nobler writing tools. Yet since way back when, now as then, the most widely distributed writing instrument in the world has gained famous fans among artists, poets and writers, as the following quotations show:

"In just this sense, I much prefer to reach for the pencil, which gives its strokes more willingly. For it often happened to me that the scratching and spattering of the pen awakened me out of my sleep-walking poetry-writing, distracted me, and destroyed a small product at its birth."

Johann Wolfgang von Goethe
from: Dichtung und Wahrheit

"I still wanted to tell you of a type of pencils by Faber that I have found. They are of this thickness, very soft, and of better quality than carpenters' pencils, really a splendid black, and one can work very nicely with them on large projects. I drew a seamstress on gray paper with one and created the effect of lithographic chalk. These pencils are set in soft wood, painted dark green on the outside, and cost 20 cents apiece."

Vincent Van Gogh
from a letter to Rappard

"For the writer of these lines there was a time when he really hated pens, was tired of them, . . . became very stupid, used them only a little, and to free himself from this excess of writing pens, began to use pencils . . . For me, the pencil helped me to play and write better. It seemed to me that the desire to write was reborn that way."

Robert Walser
from a 1927 letter

"Please, do you have 3B pencils?" "Yes, how many do you need?" "A thousand! They must be Faber-Castell No. 38," the man with the fixed pupils of the addict proclaimed triumphantly. He had to have these pencils . . . "soft enough so I don't have to press, hard enough so they don't break . . ." they had to glide softly over white paper from right to left, in Hebraic script."

Ephraim Kishon,
from a Playboy interview

After the turn of the century they dominated the office: Copying pencils of the best and finest quality, also available as pressure pencils and in pocket size, with point protector. (Faber-Castell catalog, 1906)

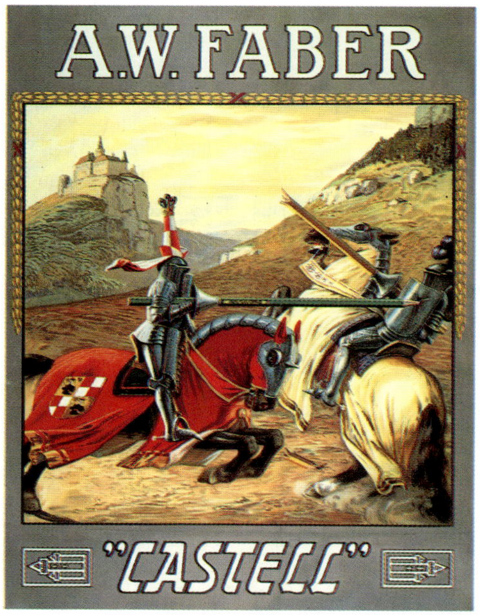

The fighting knights on the silver-colored sheet metal boxes became a symbol of quality.

Wilhelm Busch repeatedly mentioned the "Faber" pencil in his poems as a synonym for pencils in general:

Wilhelm Busch

"Meanwhile, though, Kuno draws
his Pencil Number 5 by Faber;
and Hindstroke, fussing all around,
is painfully pierced several times.
And Kuno, pleased with his victory,
Leaves the editorial office.
A real painter, smart and busy,
Always carries a sharp pencil on him."
from: Maler Klecksel (Painter Blotty)

"A little bird twitters in the twigs;
the poet feels so warm and cozy.
Gentle breezes blow around his brow;
the spirit moves him to Faber pencils."
from: Balduin Bählamm

"O Muse, hand me the pencil,
that Faber in Nürnberg must make.
Once more I'll saddle my hard trotter,
the old sticklike Pegasus!"
from: Schnurrdiburr or The Bees

"The best pencils made in Germany are still by Faber (-Castell)."
Max Liebermann

". . .the thought of the booty, the hundred confiscated pencils of the "Faber" brand still bothered me for a long time . . .
Alexander Solzhenitsyn
from: The Gulag Archipelago

Walter Kempowski
"What do I write with?" My manuscripts with a nice light, semi-soft pencil that does not break if it drops. It must be hard enough so the point does not get dull too soon, and soft enough to slow it a bit on the slightly rough paper."
from: Office Function

"Colored pencils are nice, sorted according to colors in a flat sheet metal case, and one after another. "Faber-Castell," how nice that sounds!"
from: Heartily Welcome, novel

How a Pencil is Made

To cover the precious pencil lead, wood of the Virginia cedar (Juniperius virginiana) was formerly used. It belongs to the family of juniper trees. Its wood is long-fibered, almost without branches, and very easy to cut. The red cedar required the moist soil of eastern North America as far south as Florida. The great difficulty of bringing its trunks out of the swampy interior of Florida moved the pencil manufacturers to find another kind of wood that had practically the same qualities and characteristics: the California cedar (Librocedus decurrens), usually called "incense cedar." Since it grows in the warm, dry soil of California, it is easier to cut and transport than the swamp-loving red cedar. The wood is exported from the United States in the form of small boards already cut to pencil length.

To produce pencil lead, the two natural ingredients of graphite and clay have to be purified and prepared into quality leads before being worked. The mixture is made according to precisely followed formulas; then the homogeneous mass is ground until the desired fineness of grain is attained, having a diameter of about one thousandth of a millimeter. Kneading machines and rolling mills give the material the right consistency to be pressed into plastic leads. They are baked hard in ovens at temperatures of about 1000 degrees. After baking, they are dipped and soaked in melted wax or fat, in order to increase their ability to stick, slide, and provide intensity of color.

On a fully automated pencil assembly line, the leads are laid in the grooves of the prepared boards in about thirty work processes, and covering boards are glued on. The doubled boards are cut down into individual pencils. At the end of the "pencil line" an automatic machine ejects the finished pencils at the rate of nine pieces per second! For reasons of health, high-priced pencils are dipped in paint and fitted with a cap. Marked with the brand name and grade of hardness, they are packed and ready for sale after strict quality control.

Along with the pencil, the copying instrument attained great importance in the accounting chambers of business and government. The colored pencil, invented by Jacobsen of Berlin in 1874, was equipped with the extraordinarily coloring material of aniline instead of graphite. The colored marks were very difficult to erase. By pressing on moist, transparent paper in the copying press, up to two accurate copies could be made after turning.

Manufacturing process, from board to pencil.

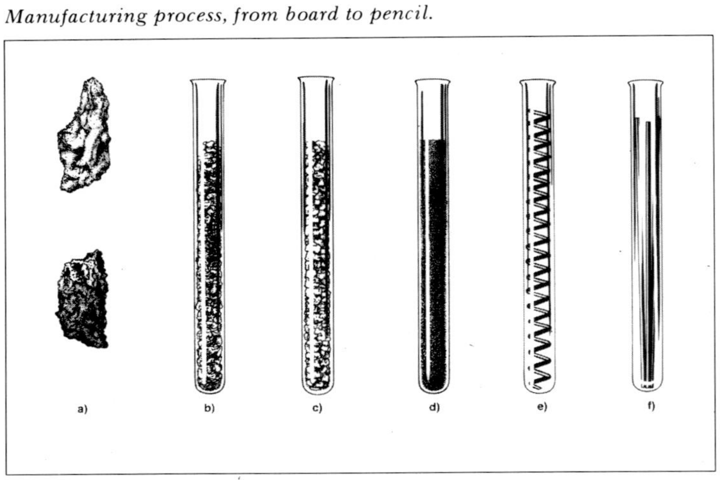

1. Preparing the lead
a) The raw materials, clay and graphite
b) pulverized clay
c) clay, purified by sifting
d) ground graphite
e) extruded, unbaked lead
f) the lead, baked and saturated with wax or fat

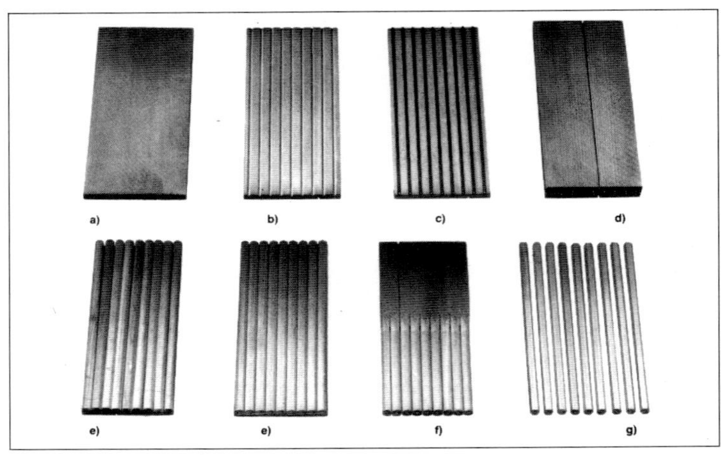

2. Preparing the raw casing
a) Raw board of incense cedarwood
b) Grooved board
c) Grooved board with leads in place
d) Doubled or glued-on board
e) Profile, round or hexagonal, planed out on one side
f) Planing the pencil profile
g) Raw pencils

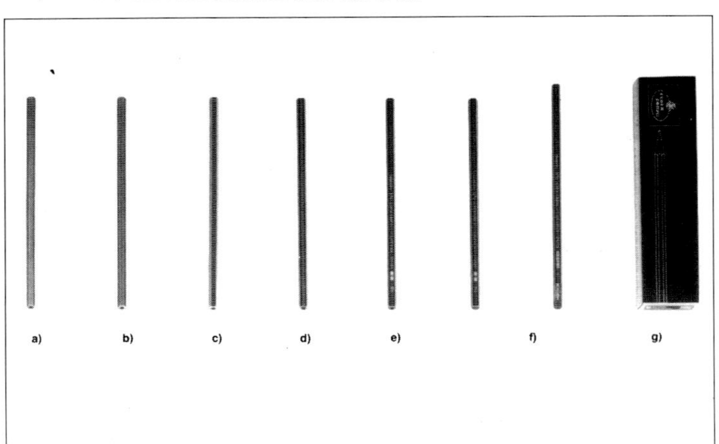

3. Preparing the end product
a) The finished raw pencil
b) The simple, polished pencil
c) The polished and striped pencil
d) Pencil with angle stripes
e) Pencil stamped, sharpened, and rounded for dipping
f) Pencil dipped and fitted with eraser cap
g) Pencils in boxes of a dozen

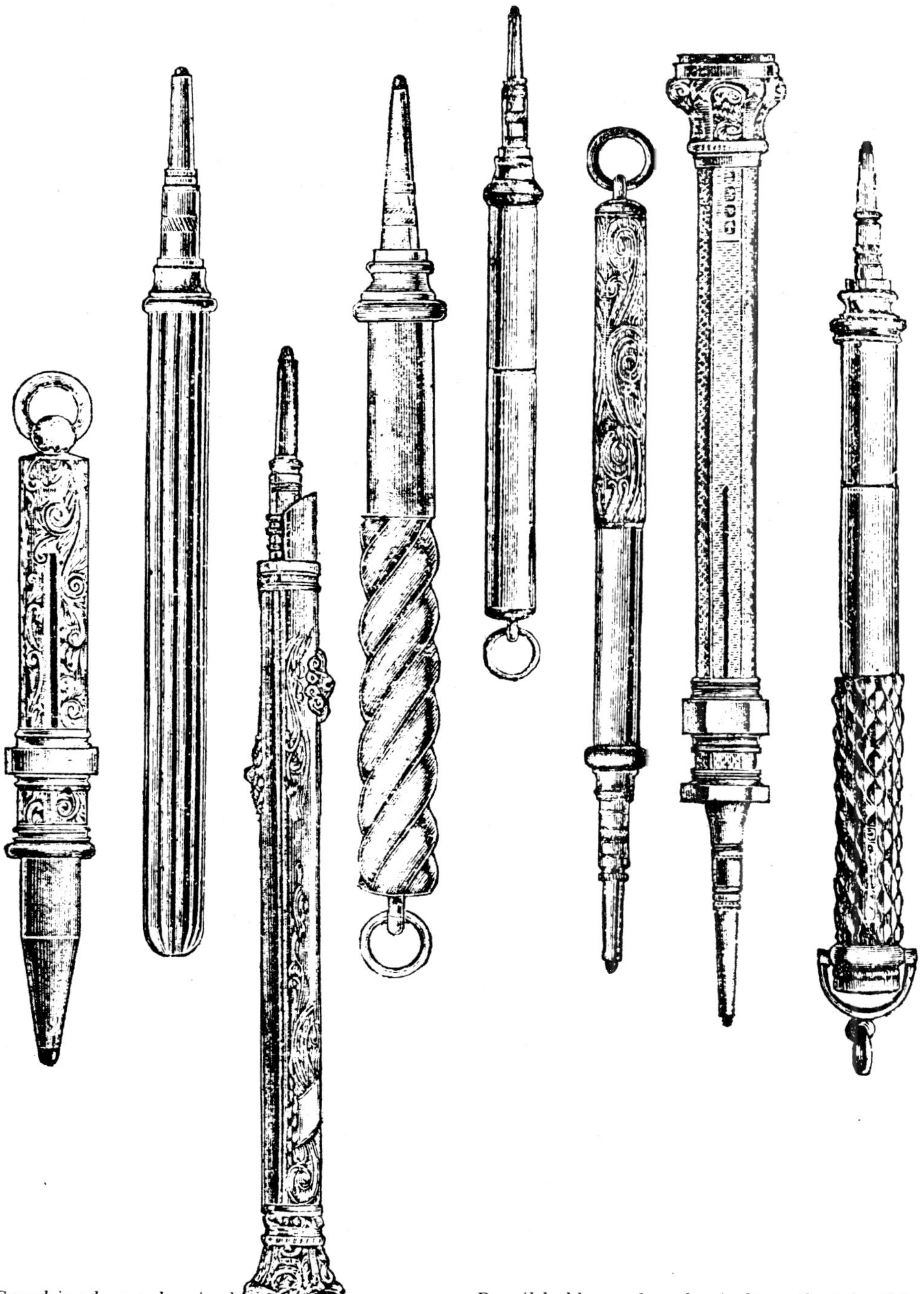

Combined mechanical pencils and fountain pens.

Pencil-holders and mechanical pencils of the 19th century.

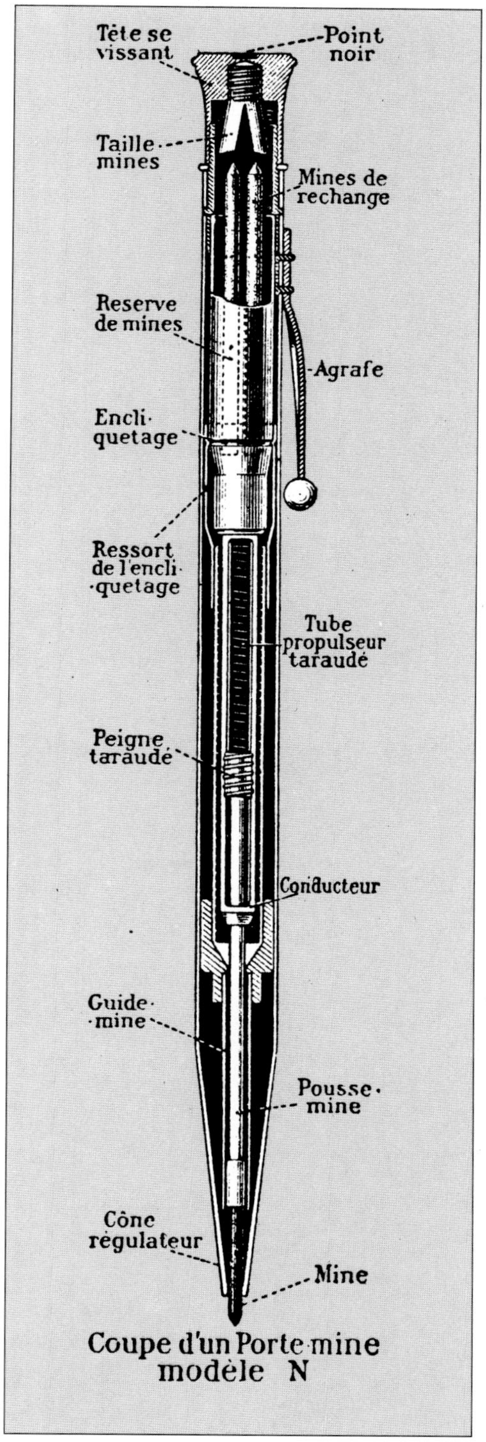

French Jif mechanical pencil of 1925.

Caricature of a mechanical-pencil mechanism by Montblanc, from the 1930s.

Faber-Castell

Lothar Faber (1817-1896), ennobled as a baron for his contributions to the Bavarian economy, gained worldwide fame for the Faber firm.

There were cabinetmakers in Nürnberg who took up "pencil making." One of them was Caspar Faber, who in 1758 married the daughter of a pencilmaker from Stein, near Nürnberg. In 1761 Caspar Faber began to make his own pencils—just as many of them as the founder of the dynasty could pack into a woven basket, carry to Nürnberg and sell to the dealers there. Caspar's son, Anton Wilhelm, gave the firm the name of A. W. Faber.

The internationally successful era in the firm's history began only in 1839, when a member of the fourth generation, Reichsrat Lothar von Faber, took over the small factory in Stein, outside the gates of Nürnberg, after the death of his father Leonhard. In Paris and London the well-trained businessman had seen how a modern business must be built up technically and economically: "Quality and nothing but quality must be the key to success." He introduced new production methods, replaced handwork by modern machines, and set up definite standards for the length, strength and grade of hardness of his pencils, standards that were soon adopted by other manufacturers and have retained their validity unchanged to this day. One of his most significant innovations was the marking of his pencils with his name. The "A. W. Faber" pencil thus became one of the first brand-name articles in Europe. To protect his products successfully from imitations, Lothar von Faber petitioned the German Reichstag to create a trademark protection law in Germany in 1874. The law went into force in 1875.

Other countries soon became important buyers of "Faber products." As early as 1859, Lothar von Faber founded the first foreign branch in New York. This American production was begun because of high import duties. His youngest brother, Eberhard Faber, became its director. Eberhard Faber later separated from the original firm of A. W. Faber, as did the middle brother Johann. Eberhard produced his own pencils in America, as did Johann in Nürnberg, in competition with their older brother Lothar. Additional business houses in Paris, London, Vienna and St. Petersburg followed. Business connections expanded as far as the Near East, and later even to China.

An important competitive advantage was gained by Lothar von Faber in 1856, when he joined forces with the geologist and businessman Johann Peter Alibert. While prospecting for gold in the Batougal Mountains of Siberia in the 1940s, he had discovered rich deposits of graphite west of Irkutsk and signed a mining treaty with the Tsar "for all time." Although the Faber graphite mine was more than 4000 kilometers from Stein as the crow flies, transporting the ore on the backs of reindeer and mules through a roadless land to the nearest port and shipping it to Stein was worth the cost. The purity of the Siberian "black gold" far exceeded that of the English graphite.

The advantage of having his own source of raw materials and being independent of suppliers' price demands also inspired Lothar von Faber to plant American cedar trees, the seeds of which he had sent to him from America, on some of his property near Nürnberg. In this way he wanted to break the American monopoly and give succeeding generations further business advantages. Unfortunately, the tree trunks, growing slowly in the Franconian climate, proved to be too hard. The last of them were cut down a hundred years later in the postwar winters following 1945 and used for firewood.

In 1870 the firm's catalog of goods included well over a hundred pages, on which the most varied writing and office utensils were offered. Most of the products were made by the company itself. Writing instruments were made in Stein near Nürnberg. Slate tablets, wooden rulers and calculating staffs from Geroldsgrün, erasers were from Newark, New Jersey, and inks and

The pencil factory at the time of its hundredth anniversary in 1861.

The first brand-name pencils, 1840.

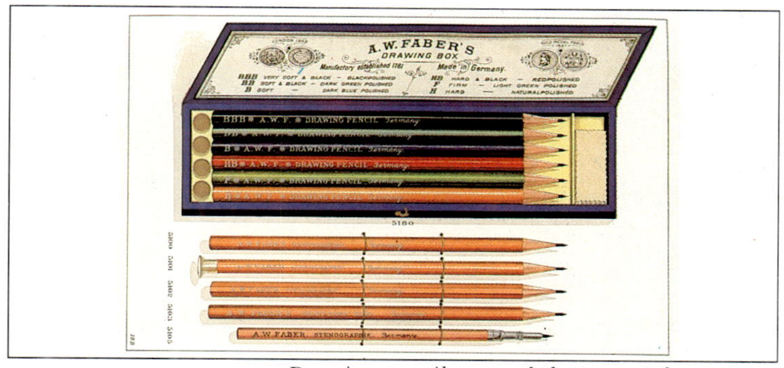

Drawing pencils around the turn of the century.

Present-day view of the factory.

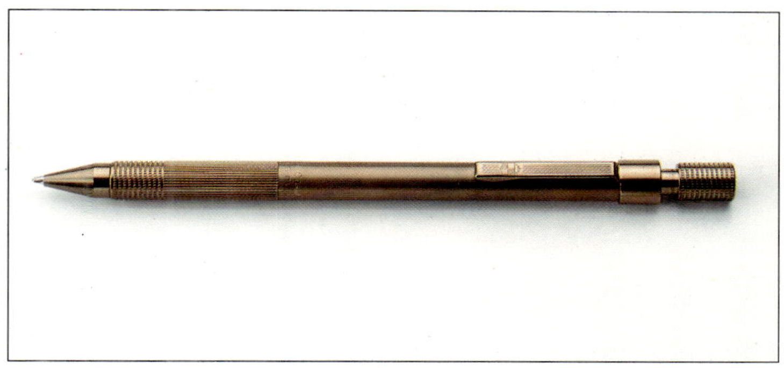

The first pressure pencil with automatic lead advance and precision writing tip.

Pencil labels for export. The exotic illustrations are still a treat today.

pens came from Noisy-le-Sec near Paris. Around the turn of the century, more than a thousand employees and "factory officials" worked for the firm of A. W. Faber.

After the death of Lothar von Faber in 1898, Count Alexander von Castell-Rüdenhausen, who had married one of Lothar's granddaughters, took over the direction of the company, which was now world famous. "A. W. Faber" now became "A. W. Faber-Castell." He expanded the factories further and became known as the creator of what is probably the most famous Faber-Castell product, the green pencil with the trade mark "A. W. Faber-Castell 9000." These pencils, first produced in 1905, bore the new double name and the two symbols: a scale (mercantile righteousness) and a castle (nobility). The expansion of the brand name was meant not only to stress the highest quality of the products, but also to set up a clear boundary between their products and the similarly named products of the competitor, Johann Faber.

In 1928 Alexander's son Roland, Count Faber-Castell, took over the leadership of the firm at the age of 23. The merger with the pencil factory of Johann Faber (Gold-Faber Series) which he achieved brought new impetus to the business in Stein.

In 1935 Faber-Castell bought the "Osmia" fountain pen factory in Dossenheim, near Heidelberg, a former branch of Parker. The firm had already made a name for itself with top-class products such as the *Osmia-Supra* piston fountain pen, but also with pressure and vacuum filling systems (*Progress*). Faber-Castell retained the brand name of Osmia and expanded the line of products with "genuine" Castell fountain pens.

In the 1960s the products were modernized. Of the former brand of Osmia there now remained only the catalog name *Progress*. The sharply increasing use in schools led to the introduction of the *Castello School Pen*, with a cartridge filling system, early in the 1960s.

World War Two brought severe losses for Faber-Castell. The export-oriented firm lost many of its foreign markets, and efforts had to be made to win them back. At times pencils had to be made with native pine wood. After 1945 the firm tried to stay afloat by producing medical instruments (the *Castalia* brand). The experience in metal-goods production gained thereby was put to use by the firm after 1945 in the development of metal mechanical pencils. The successful "TK" pressure pencil for technical work won world fame after 1948. In 1949 Faber-Castell was the first German pencil manufacturer to produce ballpoint pens. Early in the 1960s the firm began to produce its own products to compete with Japanese coloring instruments, the felt- and fabric-tip markers.

In 1978 Count Anton Wolfgang von Faber-Castell, as the sole executive officer, took over the leadership of the firm from his father Roland. With this change of leadership came a "cleaning out" of production and concentration on modern products for technical drawing, artistic and office use. The mechanical drawing pencil factory in Konstanz and the fountain pen factory in Dossenheim were given up in 1975. New products such as fine polymer leads, the first fully automatic fine-lead pencil in the world (TK-matic), modern ink rollers, fine-tip writers in a high-quality artistic assortment, as well as prestige writing instruments to Porsche designs hand-finished in silver and titanium are the jewels of a new and impressive series of writing instruments made by this German pencilmaker. Since 1978 the firm has also produced high-quality cosmetic cases for internationally known cosmetic manufacturers.

In 1986 this tradition-rich firm, with its 400 employees worldwide, celebrated its 225th anniversary. Faber-Castell may well be one of the oldest industrial firms in the world that, now as before, has remained in the same family. The company, now directed by the eighth generation, ranks among

Illustrations on page 57, above:
A. W. Faber pencils, pencil lengtheners and protective caps won world fame.

Illustrations on page 57, bottom:
Along with lead and copying pencils, A. W. Faber also sold fountain pens and ink, which were produced in Noisy-le-Sec, near Paris. Photos from the 1900 catalog.

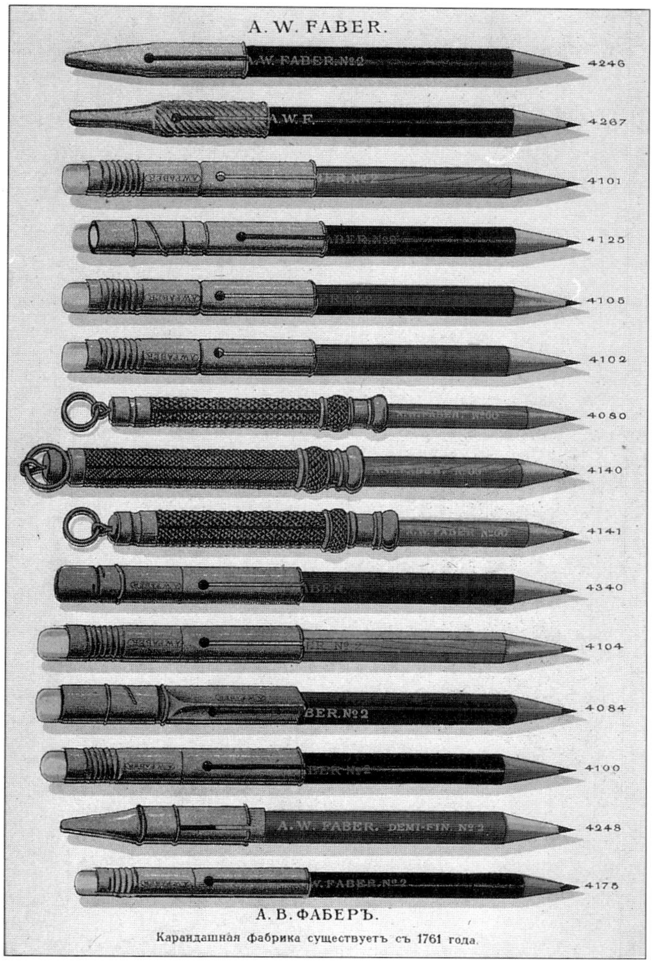
A. W. FABER.
4246
4267
4101
4125
4105
4102
4080
4140
4141
4340
4104
4084
4100
A. W. FABER. DEMI-FIN. №2
4248
4175
А. В. ФАБЕРЪ.
Карандашная фабрика существуетъ съ 1761 года.

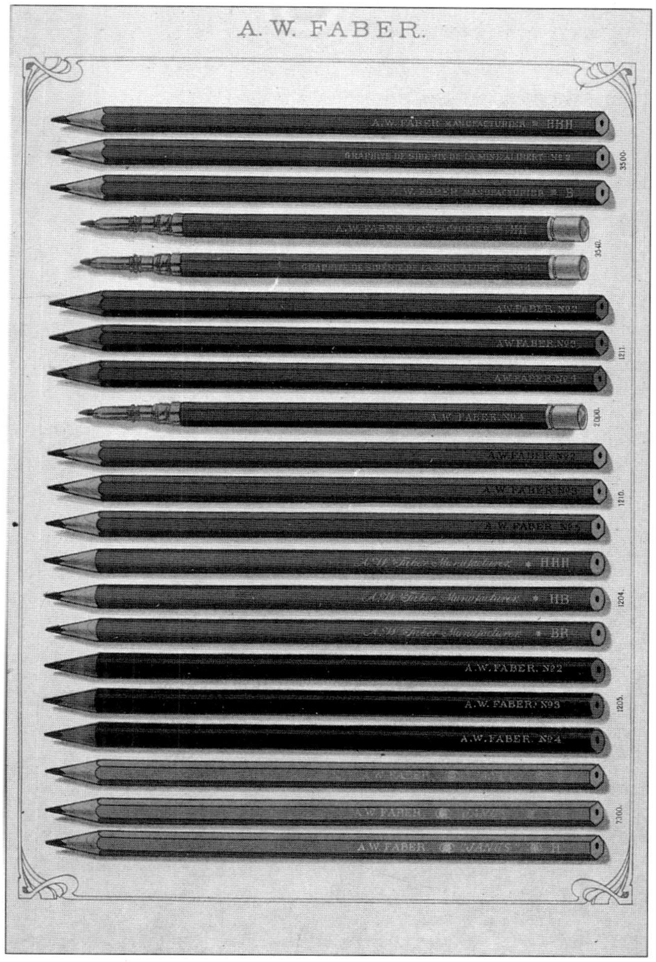
A. W. FABER.

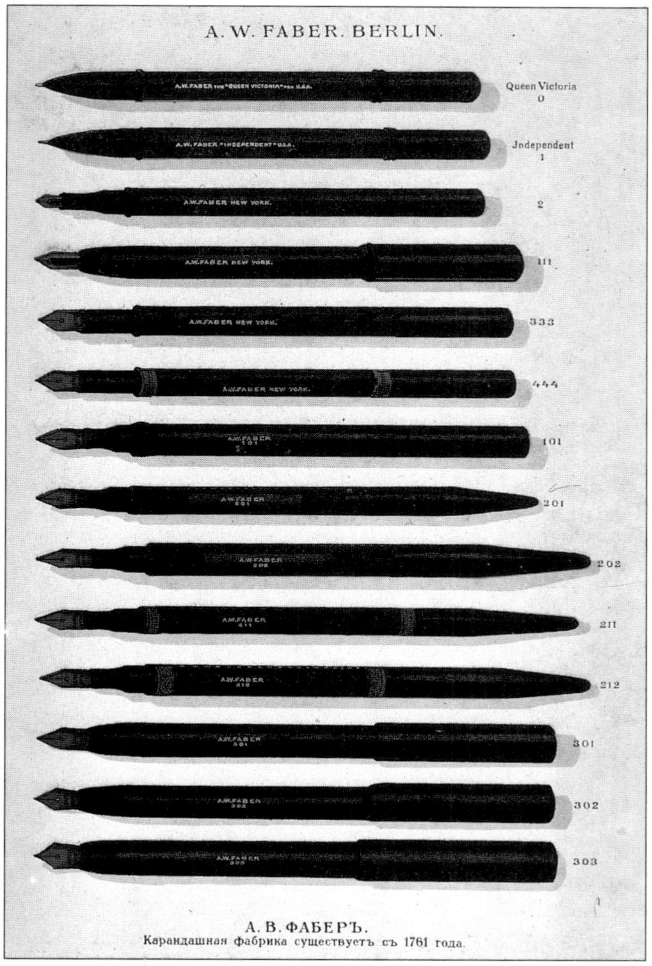
A. W. FABER. BERLIN.
Queen Victoria
0
Independent
1
2
111
333
444
101
201
202
211
212
301
302
303
А. В. ФАБЕРЪ.
Карандашная фабрика существуетъ съ 1761 года.

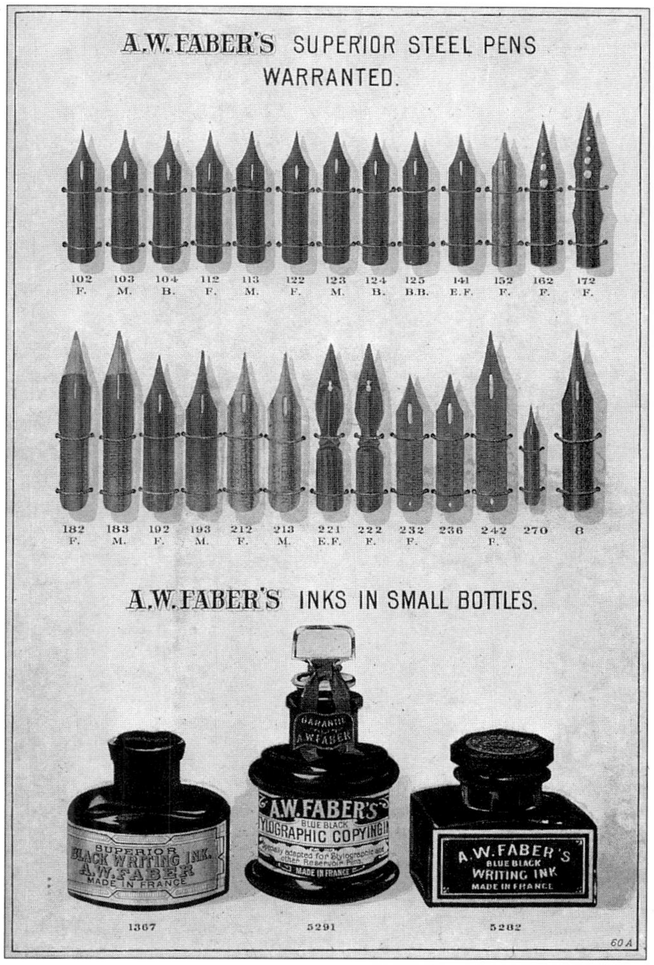
A.W. FABER'S SUPERIOR STEEL PENS
WARRANTED.
102 F.
103 M.
104 B.
112 F.
113 M.
122 F.
123 M.
124 B.
125 B.B.
141 E.F.
152 F.
162 F.
172 F.
182 F.
183 M.
192 F.
193 M.
212 F.
213 M.
221 E.F.
222 F.
232 F.
236
242 F.
270
8
A.W. FABER'S INKS IN SMALL BOTTLES.
SUPERIOR
BLACK WRITING INK.
A.W. FABER
MADE IN FRANCE
A.W. FABER'S
BLUE BLACK
STYLOGRAPHIC COPYING INK
A.W. FABER'S
BLUE BLACK
WRITING INK
MADE IN FRANCE
1367
5291
5282
60 A

the international leaders in its field, with more than two thousand products. What did Lothar von Faber write in his diary in 1855? "From the beginning I always had the intention of raising myself to first place by making the best that is made in the world . . ."

J. S. Staedtler

The founding of the firm of J.S. Staedtler coincides with an important date in German industrial history. In 1835 the first railroad line was opened from Nürnberg to Fürth. The name of Staedtler, though, had already been linked with that of the pencil for almost two hundred years.

In seventeenth century Nürnberg the work of making pencils was divided. "Messerschalen-Schroder," who cut graphite from English mines with special saws, delivered leads to carpenters, boxmakers, and cabinetmakers. The woodworking artisans planed deep grooves into pieces of wood, inserted the leads and closed the open grooves with small pieces of wood. The whole thing was then rounded. Soon enough, though, a Nürnberg man who did not belong to the cabinetmakers' guild got the idea of earning his daily bread by making pencils. On January 7, 1662 the merchant Friedrich Staedtler asked the Nürnberg city council to allow him to make lead-white pencils. His petition was turned down, but Staedtler must have persevered with his idea of combining the cutting of graphite leads with the woodworking at his workshop. After 1662, church records always list him as a pencilmaker. His sons and succeeding generations remained loyal to their ancestor's trade and made pencils as the centuries passed, along with their apprentices and employees, and with the help of family members too.

Then times became worse. The Cumberland graphite mines were exhausted; the ceramic process of making pencil leads, developed at the end of the eighteenth century, gave France and Austria an impermanent monopoly in the pencil trade. The Nürnberg pencilmakers were close to ruin. Finally Paulus Staedtler, a great-grandson of Friedrich Staedtler, tried a new technique. He was lucky, and thus a member of the old pencilmaking family laid the groundwork for a new upswing in the old free city. Down-to-earth artisans took a chance and became manufacturers.

At the same time as this industrial breakthrough, the firm of "J.S. Staedtler" was founded by a descendant of Friedrich Staedtler. On October 3, 1853 he was granted a concession to build a pencil factory. It was equipped with graphite mills, cutting, packing and rounding machines and baking ovens. By 1875 it already employed 54 workers and had achieved a production of two million pencils per year. Cinnabar and other coloring materials were used in the production of pencils that soon became a specialty of the house of Staedtler known far beyond the boundaries of Germany.

The first hand-carved pencils from the Staedtler archives.

At the Nürnberg Industrial Exposition of 1840, the firm displayed 63 different types of pencils. Their sales areas included the states of the German Customs Union, Austria, France, England, Italy, Russia, and the Orient. Traveling salesmen who brought in orders for numerous firms were already underway throughout the world at that time.

During the 1870s, the pencil factory of J. S. Staedtler had to struggle with financial difficulties. In 1880 the Nürnberg businessman Ludwig Kreutzer joined the firm and became its sole proprietor. Along with his brothers, he was able to bring new life to the firm and build it into a business with a worldwide reputation. The firm's trade mark at that time was a half-moon, which had been registered with the government patent office in 1894. In East Asia it was called the Moon Pencil Factory. Along with this, *Luna* and *Camel* developed into world-renowned brand names of pencils from Nürnberg. At the turn of the century, the brand name of *Mars,* still well-

Staedtler pencils of 1908 ▷

The Mars brand was the highest-quality line of Staedtler pencil production. The half-gross package of Mars copying pencils dates from 1935.

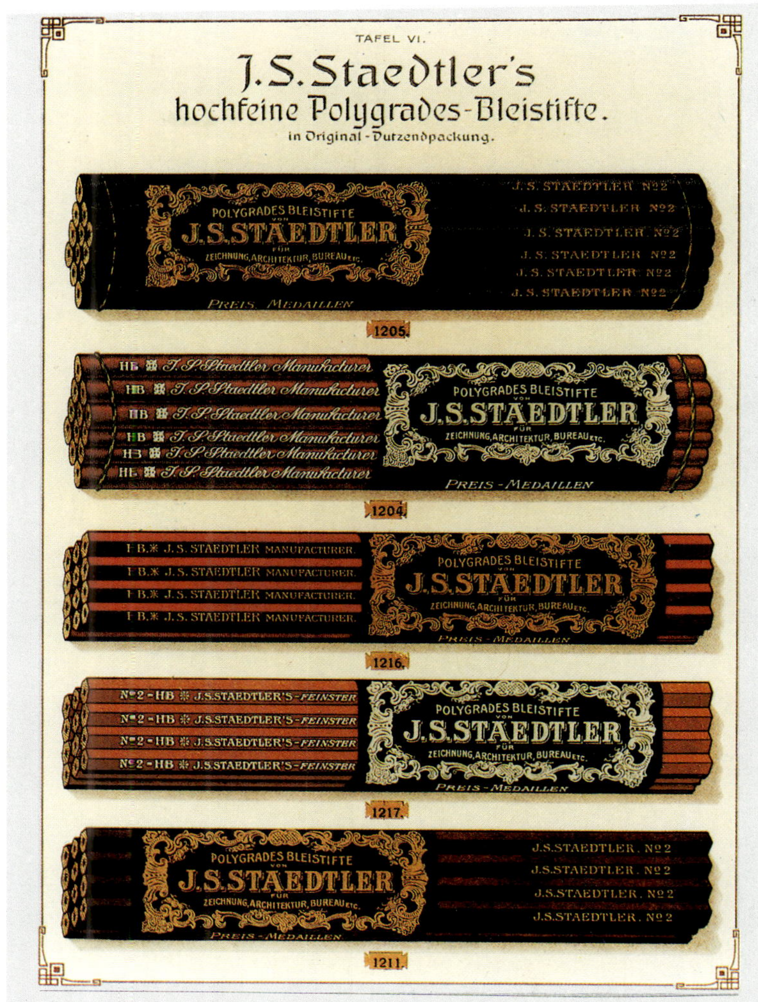

Samples of Staedtler fountain pens available in 1955.

known today, was registered for the Staedtler firm's highest-quality pencils. The "Mars-blue" color soon became the identifying mark of the *Mars* copying pencils that, along with the brand of *Noris*, were found in offices and chancelleries all over the world.

The firm achieved decisive progress after the young Dr. Rudolf Kreutzer, a son of Ludwig Kreutzer, entered the Nürnberg firm in 1905. On extended travels through Asia and North America he studied the sales markets thoroughly, built up a sensible assortment of products, and preserved the reputation of the enduring trade names.

The lead and color pencil factory of J. S. Staedtler, which had produced nothing else until then, expanded its production program in 1937. Until that time, mechanical pencils had only been handled as trade articles. After the purchase of a small fountain pen and mechanical pencil factory, the firm, now renamed the J. S. Staedtler, Mars Pencil and Fountain Pen Factory, now produced a complete line of writing implements itself.

World War Two interrupted the firm's vigorous development. Parts of the factory building and the production facilities were reduced to rubble and ashes. Only in 1946 could production begin again on a modest scale. That revolutionary new writing instrument, the ballpoint pen, which American soldiers soon made popular in Germany, also attracted the interest of the old established Nürnberg pencil and fountain pen firm. Its own production began in 1949.

In 1950 the firm's ownership changed. The majority passed from the owners to the Rudolf and Clara Kreutzer Foundation. In this way the Rudolf Steiner school system and the anthroposophically oriented sciences became the beneficiaries of the firm's profits.

In 1950 the head of Mars, which had been a symbol of the highest-quality *Mars* pencils since the 1930s along with the half-moon, became the factory's trade mark. In 1964 Staedtler was one of the first German manufacturers to expand its sales program to include felt- and fiber-tipped pens as well as overhead projection products. Synthetic erasers were produced as well. In 1962 the Mars 500 piston-operated drawing-ink fountain pen appeared, wearing the well-known Mars head symbol. In 1971 the technical drawing program completed a broad assortment of cutting instruments. As of 1967, Staedtler ranked among the few firms in the world that had built up their own high polymer fine lead production facilities.

The firm of Eberhard Faber in Neumarkt was taken over in 1978. With its products selling in every part of the world and a worldwide sales and delivery network, Staedtler has developed from its modest beginnings as a "pencilmaker" into one of the leading present day producers of writing and drawing instruments.

Pencil labels for export: The Luna and Camel brands developed around the turn of the century into world-famous brands of German pencils made in Nürnberg.

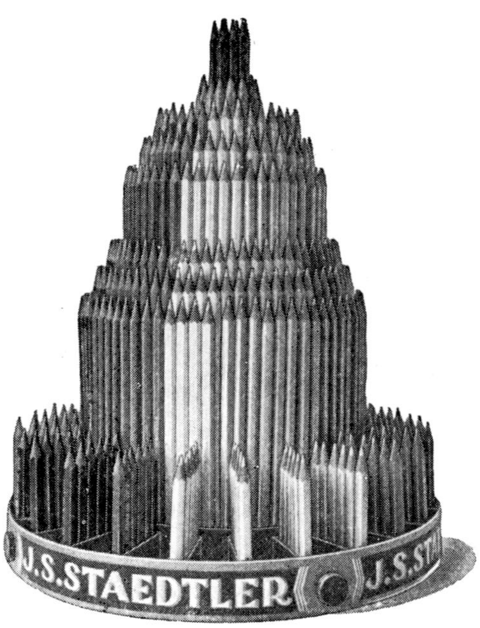

An attention-getting show-window decoration.

Along with their own lead and colored pencils, Staedtler also sold a wide range of mechanical pencils and fountain pens (piston, pipette, pressure, and lever filling systems). Title page of the 1935 catalog.

MARS
The victory of perfection!
In its styling it presents the most modern type of the ideal fountain pen. Quality and durability, elegant decoration and solid workmanship, all united in one. A companion for life, of lasting value, and yet easy to obtain.

Schwan-Stabilo

Graphite mill, 1859.

Around the turn of the century, Schwan literally wrote its advertising in big letters. In 1906 the Schwan Pencil Factory erected this 30-meter-high giant pencil in Nürnberg. To this day, it is the world's largest pencil.

"4 lead-white mills, which are run by horsepower, plus 1 circular saw, 2 cutting machines, 1 lead press, 1 baking oven and 1 etching kettle" were the technical equipment with which the businessmen Georg Grossberger and Hermann Kurz set up a modern pencil factory in Nürnberg in 1854. On a piece of land not far from the imperial castle, 25 to 30 employees made pencils of extraordinary quality. Modern production methods tall smokestack, indicating the use of steam power.

In 1865 Gustav Schwanhäusser took over the business. The 25-year-old came from a respected bourgeois family which took its name from the town of Schwanhausen on the boundary between Franconia and Thuringia. In reference to his name, he used the figure of a swan as the identifying mark of his pencils and the trade mark of his firm, which was known from then on as the "Schwan Pencil Factory, Schwanhäusser, formerly Grossberger & Kurz," and outside Germany as the "Swan Pencil Co."

Under Gustav Schwanhäusser the company grew quickly. At numerous expositions on all five continents the Swan pencils won gold medals and first prizes. The organizational expenses for exporting to many countries must have been imposing. The sales network was still in its first stages. The telephone and typewriter had not yet been invented. Letters were written by hand with a pen or copying pencil, which got its name because one could make copies of its violet aniline pigment on the copying press. Gustav Schwanhäusser, the first manufacturer of this pencil, recognized its advantages and obtained an imperial patent on it in 1875.

Talent for mercantile and technical success passed from the father to his sons. Before Dr. Eduard Schwanhäusser took over the business end of the firm in 1893, at the age of twenty-two he had already written his doctoral dissertation on the subject of "The Nürnberg Pencil Industry." It is still regarded today as the standard work on the history of pencil manufacturing. August Schwanhäusser devoted himself, as a chemist, to the direction of the firm's technical department. In the following years he made a number of important inventions and improvements. Until then, if a colored pencil was to draw a strong line, a thick lead of mineral, vegetable, or animal coloring agents had to be used. In 1925 Schwan introduced the first thin-lead coloring pencil, the *Stabilo*, which far exceeded anything previous in terms of coloring power, endurance, and even hardness. In 1927 came the development of the world's first wooden cosmetic pencil, an eyebrow pencil produced by Schwan. In 1930 Schwan produced the first fully weatherproof copying pencil, the *Cellomin*, in 1932 the paintable coloring pencil (*Swano*).

On January 2, 1945 the Schwan factory was destroyed to the bare walls. Rebuilding began in May of 1946. At first cosmetic products such as shaving lather and lipstick were produced, and later lead and colored pencils were added. In 1956, one year after the firm's hundredth anniversary, came the invention of the new *All-Stabilo* graphite pencil, that could write on any flat surface. It gave a new impetus to the sale of wooden pencils, which was becoming more difficult.

Under the leadership of Günter, Otto, and later Horst Schwanhäusser, the 1960s and 1970s led to a large-scale expansion of Schwan production. In 1966 the first pointed fiber *Stabilo-Pen* appeared, followed,in 1967, by the *Othello Kugler 5 km* ballpoint pen. In the same year, part of the Soennecken firm were bought. In 1968 Schwan-Stabilo put its first products for modern overhead projection on the market: fiber-tip writers, foil and other accessories. In 1971 the firm came up with a surprise in pioneering marketing. The introduction of the *Stabilo Boss* text-marking pen, with a newly patented vivid marking ink that made it possible to mark important passages with glowing clarity simply by drawing it over them. The *Boss* became one of Schwan-Stabilo's most important products in the years that followed.

In 1972 production of a new wooden-pencil cosmetic series began In all the world there was scarcely a single leading cosmetic firm that was not supplied by the firm of Schwan. The names of Revlon, Elizabeth Arden, and Dior are only a few among many. Contact with the international cosmetic centers of Paris, New York, Tokyo, and London is only natural for the Nürnberg firm.

Today Schwan-Stabilo ranks as the leading international manufacturer of markers and cosmetic pencils, produced by some two thousand employees in Nürnberg, Weissendorf in Bavaria, Mexico, and Malaysia.

Advertisement for children's drawing pencils, from the 1920s.

Advertisement for the weathertight copier pencil, 1930.

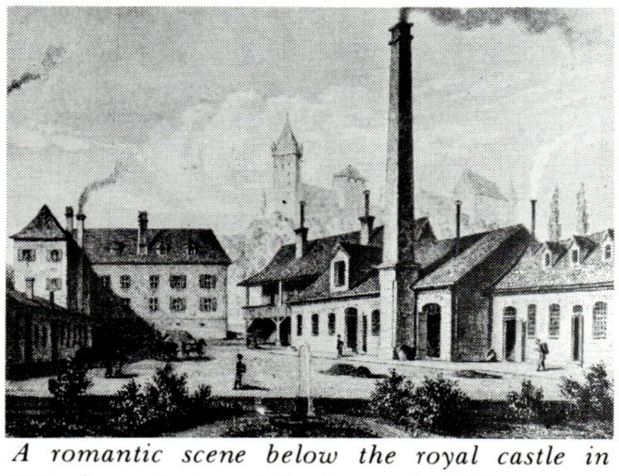

A romantic scene below the royal castle in Nürnberg: A view of the factory in 1860.

The Fountain Pen—

"Of fountain pens it may be said, as of other things, that only the most useful survive."
P. Maginnis, in "Royal Society of Arts," 1905.

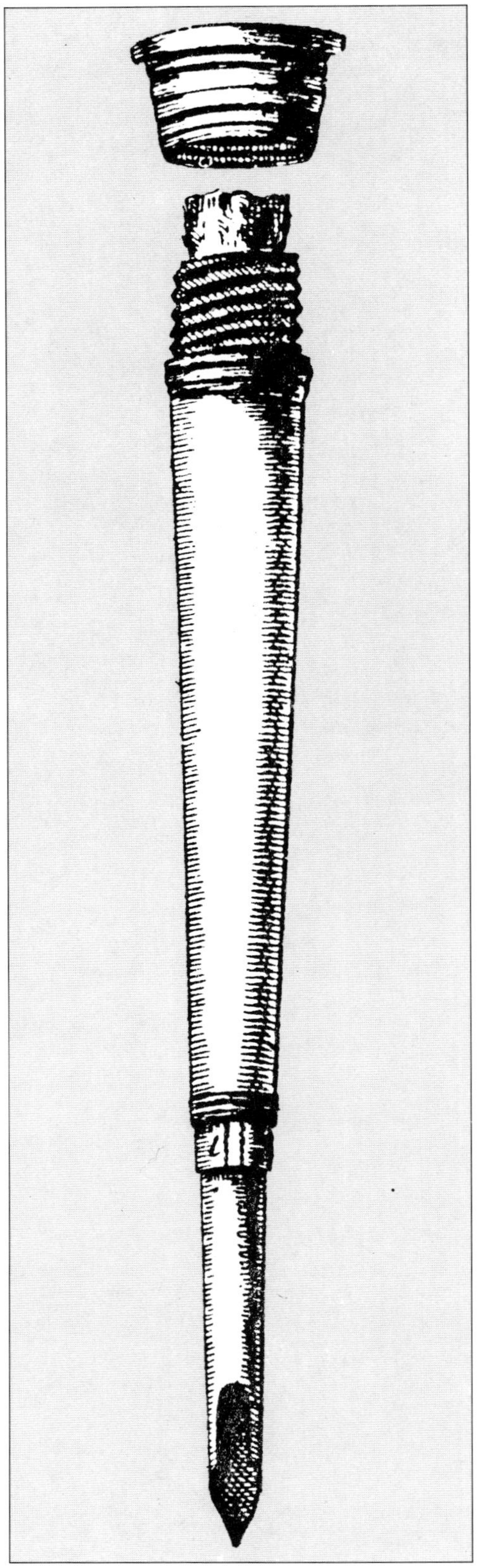

Portable pen with ink, from 1781.

The understandable wish to do without that laborious dipping into the inkwell and write with an ever-filled pen is as old as writing with pen and ink itself. And naturally, only people in high positions had the chance to make this wish come true.

Probably the first recorded attempt to invent a pen "that carries its own ink supply with it and only lets the ink flow when the pen is used for writing. (A pen) which one can carry along in one's sleeve at other times without having to worry about one's clothing being stained," is described by Albertine Gaur. In his book, *Writing Materials of the East,* he tells of the Caliph al-Múizz, who issued a command to his court artisans in the tenth century. After long, laborious experiments, the masters, as fearful as they were capable, were able to place such a pen of pure gold in the soft white hand of the Moslem ruler.

Nothing of the way this golden pen functioned is told, nor is it in the report of the pastor Johann Mathesius, from the Bohemian silver-mining town of Johannisthal (from which name that of the "Thaler" coin and in turn that of the American

Ein Federn zuschneiden/daß sie nicht klecke.

Wann die Feder abgeschnidten vnd auffgespalten/vnd das vorder theil zum schreiben tüchtig / so schneide das vntere Spältlein nicht herauß / wie man sonsten pflegt/damit schreibe/so wird sie keinen Dalcken werffen/dann auff diese Manier kan sie nicht mehr Dinten fassen / als sich gebühret / so aber das vntere Spältlein abgeschnidten / vnd man zu hart einduncket/ fasset der raum vnter dem ort/davon das Spältlein geschnidten / zu viel Dinten/ welche hernach / wann man nicht vorsichtig handelt / fallen/ vnd das Papier beklecken muß.

Vierzehender Theil der Erquickstunden.

ein stück/ wie 2/5. schneide bey 4 ein löchlein darein / daß vngefähr ein Hirschkörnlein dadurch fallen könne / die stuck stecke zu oberst auff die drey stück/ daß das löchlein oben her komme / solches versteck mit einem stuck Kyhls / wie bey 9/5 zusehen/vnd vom fünfften Kyhl geschnidten ist / vnnd dann von dem sechsten Kyhl præparire eine gute Schreibfeder/ schneide das hole ab/vnd spalts in der mitte voneinander / so wirds wie 6/7. Letzlich schneide vom siebenden Kihl ein stuck / wie 3/5. mit ein löchlein 4. legs in das halbe Röhrlein 6/7. daß das löchlein 4 vntersich komme / solchs stück 6/7. sampt dem stücklein 8 so darinn ligt / muß vnten in das vntere theil der Federn gestossen werden/daß das stuck x. 7. vor stehe/ so ist die Feder bereitet.

Solche nun zu gebrauchen / thue oben das Hütlein ab/ steck sie vnten bey 7 in eine Dinten / so fein sauber / rein vnnd dünn/nimb die Feder oben in den Mund/ziehe den Odem an dich/so wird sich die Feder voll Dinten ziehen / setze das Hütlein oben wider fein geheb darüber/so wird nichts außlauffen/ damit das vacuum verhütet werde / So man aber die Feder vnten bey dem 8 drucket / bekommet die Dinten an zweyen orten Lufft/vnd kommet ein tropffen herfür/ den kanst du verschreiben/vnd hernach so offt drucken/ so offt du mehr Dinten von nöthen. Ist einem Studioso oder Landmesser eine sehr nützliche Feder.

32

A "very useful pen," with an ink supply, was described by Professor Schwentner in his book, "Deliciae Physico Mathematicae", published in Nürnberg in 1636.

of the Early Years

"Dollar" are derived) dating from 1538. He is said to have succeeded in inventing "writing pens made of sheet metal, which you can carry with you full of ink."

A century later a mighty military leader, more at home with swords and lances than with "pen wars," had the pleasure of holding one of the world's first fountain pens in his mighty hand. In 1632 the city of Augsburg gave the Swedish king Gustav II Adolf not only an artistically carved and turned cabinet but also an equally artistically worked "ink container." Its appearance was described as follows: "At the end is a point, onto which the button is screwed above. ink is poured in and filled full . . . below, at the fourth container, the tube is removed and the silver pen-point or a properly cut quill pen is inserted . . . thus the ink runs down through the silver tube and can write for one hour or two . . ."

This was obviously an unscrewable tube with a kind of filler cap, into which a metal or goose-quill pen could be inserted.

A mathematics professor from Altdorf took up the Augsburg idea in much less complicated form. Daniel Schwentner's "beautiful secret of making a pen that

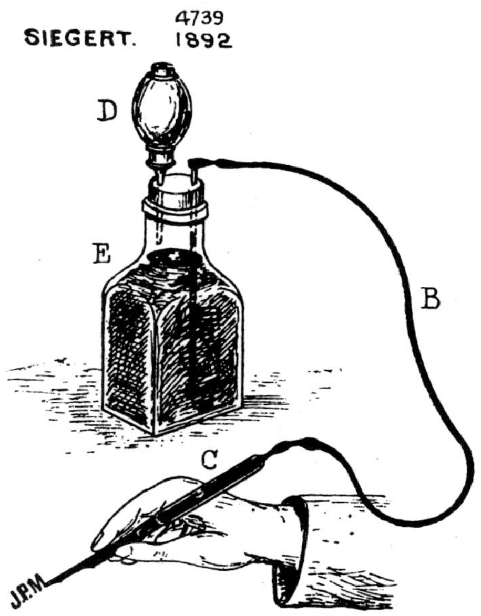

In 1892 H. Siegert suggestedconnecting ink bottle and pen with a thin rubber hose that was immersed deep in the ink. When one pressed the rubber ball (D), the air pressure in the bottle was to push as much ink into the pen as the user's writing speed required. If use in larger offices required, several people could be connected to one ink bottle at the same time.

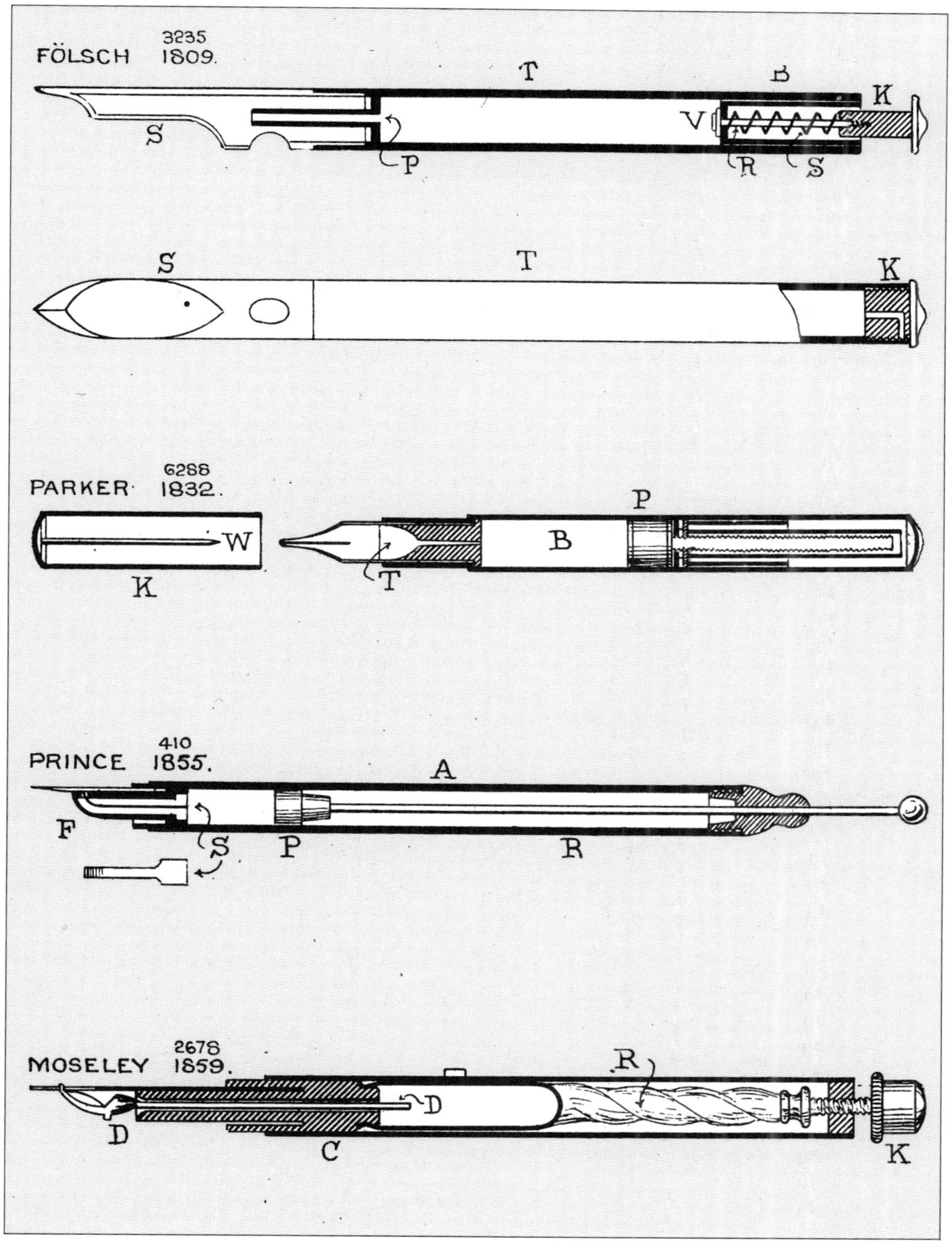

Hundreds of ideas for a way to fill a fountain pen with ink were thought up by inventors during the 19th century. Although none of the pens described here was actually made, some examples anticipated successful filling principles: the valve (Fölsch), screw piston (Parker), pressure piston (Prince), and tube filling (Moseley).

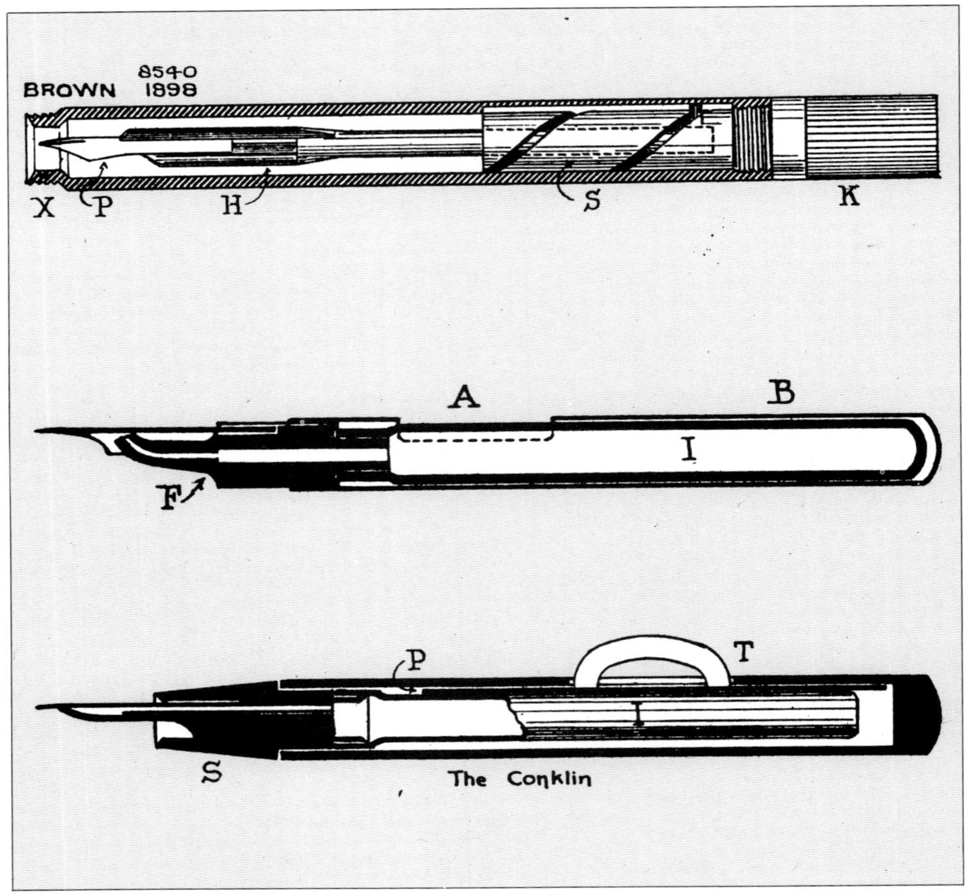

Early designs for useful fountain pen mechanisms:

Above: Safety filler with unscrewable nib.

Center: Tube filler. Pressing a finger on the opening (A) compresses the tube inside.

Below: Conklin tube filler, filled by pressing on the metal ring (T).

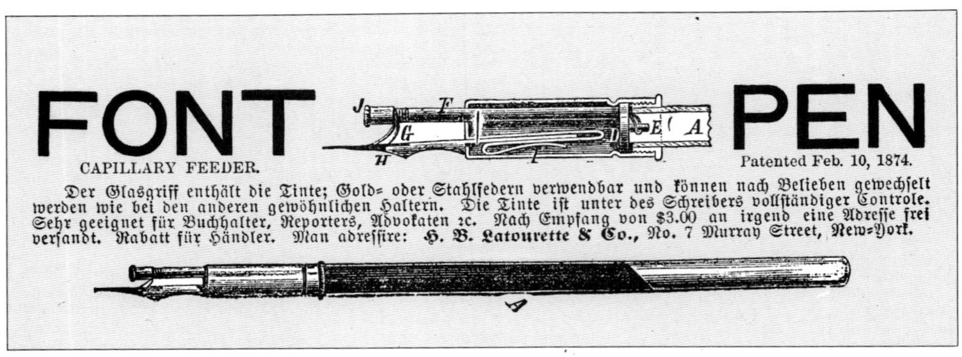

A really unusual system is this fountain pen with a glass grip.

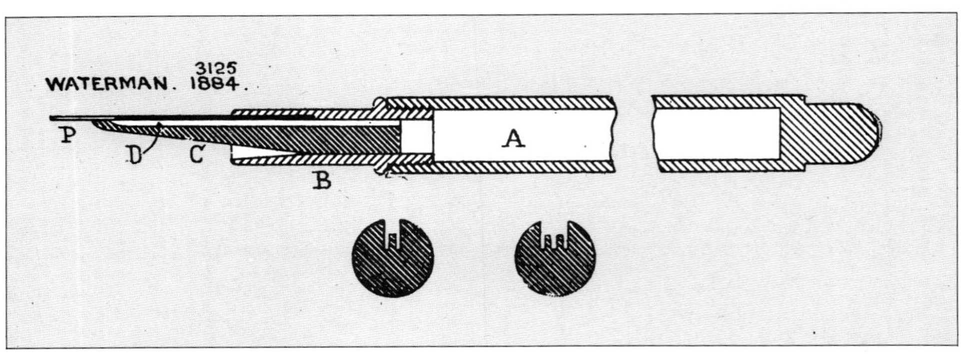

The first Waterman fountain pen. The cutaway ink passage design shows the principle of capillary action for ink and air supply.

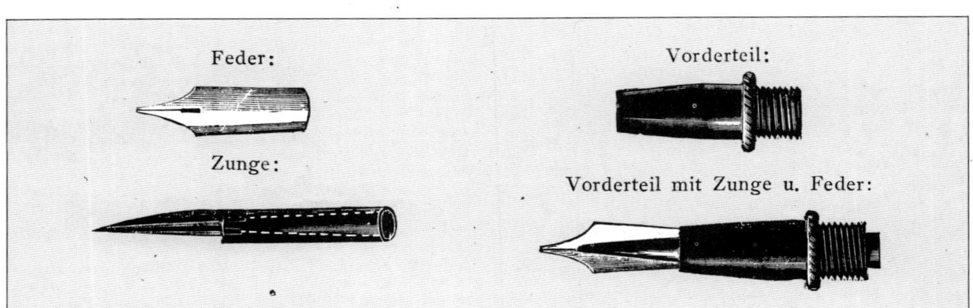

Feder: nib
Vorderteil: front section
Zunge: tongue
Vorderteil mit Zunge und Feder: front section with tongue and nib

Components of the Soennecken gold fountain pen.

holds ink and lets out as much as is needed," dating to 1636, existed only in the form of a woodcut on paper, but its simple design, consisting of three goose quills, one inside another, was impressive. The simple quill pen had a hollow intermediate section with a tiny hole in its closed bottom. When one filled this section with ink, it dripped slowly through the hole to the pen-point. The good professor anticipated the ink-supply principle of the later fountain pen, at any rate.

It took a good 150 years until another inventive brain worked on the idea of the filled pen. In 1780 the mechanic Scheller of Leipzig, intrigued by the thought of a "nice traveling pen," is said to have developed one pen model each of bronze and horn. He too used the time-honored goose quill as a pen, "which constantly has ink in it, and can be carried in the pocket . . . If one often wanted to write something down in taverns, time and mood passed before one could get ink and pen . . . By means of this pen, one can use every moment . . . since writing with a pencil soon is rubbed out . . ."

At the beginning of the nineteenth century, the demand for a functioning filler pen became more and more urgent. But oddly enough, the inventors and improvers of the fountain pen were less concerned with problems of even ink flow than with the filling, the supplying of the pen with ink, and the closing of the instrument, which long remained the experimental realm of countless trials and errors. And it was mainly inventors in Great Britain, in the USA, and in Germany as well, who worked on ink-filled pens. In 1809 F. B. Fölsch obtained the first patent for a fountain pen in England. Fölsch's fillable pen was equipped with a filler cap at the end of the barrel. To use the pen, one opened the cap, for when closed, it prevented ink from running out when the pen was not in use. The oldest German patent for a fountain pen was obtained in 1878. In July of 1827 a certain George Pulton designed a "self-supplying pen," made of a cylindrical steel tube. A weight inside the tube pressed the ink toward the nib.

John J. Parker (not to be confused with the famous George S. Parker, of whom a thorough account will be given later) created the forerunner of the present-day piston fountain pen in 1832. When one turned the filler shaft, the piston pushed the ink through a narrow channel to the pen's nib. When one pushed the piston back, the pen could be filled. A fine wire, fastened concentrically in the cap, prevented the ink from running out during transport. When one put the cap on the pen, this wire pushed into the narrow ink passage between the ink reservoir and the nib and closed it. The slightest bend in the wire, however, turned the pen into an instrument of mess-making.

In 1855 Newell A. Prince developed a fountain pen made of hard rubber. To conduct the ink, Mr. Prince used a flat piece of metal that picked up the movements of the nib and thus brought ink from the reservoir to the nib. The pen was filled with ink by using a simply piston rod, as is done in the familiar principle of the sprayer or air pump

A "hidden" solution was provided by Walter Moseley in 1859, when he put the flexibility of a tubular rubber sack to good use. To fill his fountain pen with ink, he turned a knob on the end of the barrel a few times, so that the sack was twisted and the air expelled. Then he dipped the pen in ink and let the button go, and the ink sack returned to its original form and filled with ink. Forty-four years later, L.E. Waterman was to return to this filling principle. In 1892 H. Siegert got the idea of increasing the ink supply. His "revolutionary" idea consisted of not putting an ink reservoir in his fountain pen. This idea had the added advantage of preventing the ink supply from running out. S. Fox made the pen barrel of flexible material and connected it, by means of a thin rubber hose, to an ink bottle that had to be higher than the writing hand, in a cabinet, for example, or even in the writer's jacket pocket. The hose was to run through a buttonhole or around the writer's neck.

In October of 1883, Robert Shaw patented a further curiosity. The ink reservoir of his fountain pen was to be filled with long, thin wires, bristles, hairs or other fibers that reached to the underside of the pen. When one saturated the bundle of fibers with ink, capillary attraction held the drops of liquid and gave them off to the pen

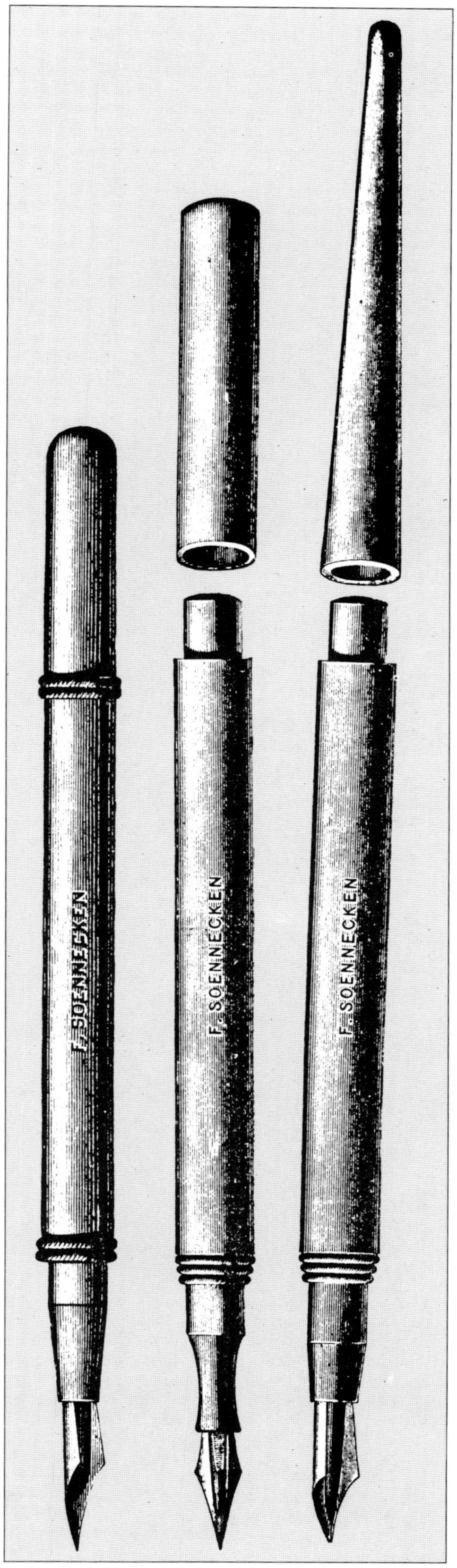

The first German insertion filler of 1890 resembles the Waterman to a remarkable degree.

in small doses. Spattering was practically ruled out. Robert Shaw certainly never knew that he had anticipated the principle of the modern fiber pen.

In some cases the "filler pens" described here were made public only in patent documents but never actually produced and put on sale. Others presented a functioning principle that would be developed further and used in fountain pens all over the world in the early years of the twentieth century. What was praised by countless inventors as a novelty toward the end of the nineteenth century could scarcely attract a broad public. The designs were too unreliable. It happened too often that the pens failed to do their job when needed or spewed out ink just when it was not wanted, often making their owners' pockets overflow regularly with blue or black ink.

The Fountain Pen Pioneer: L. E. Waterman

Thus the fountain pen soon acquired a bad reputation. This can be illustrated by a good example. It was 1885 in New York. For the 45-year-old insurance agent Lewis Edson Waterman, business success depended on the reliability of pen and ink when he needed to have his clients sign a contract on the spot. To be ready for any eventuality, he always carried a pen and an ink bottle with him in a special case.

One day the contract for a big life insurance policy was about to be signed. In order to be able to endorse the contract gracefully with the name of the insured party, Waterman decided to use a new fountain pen for the occasion. The modern device, which did not need to be dipped into the ink bottle, would surely impress his client highly. The contract lay on the table, completely filled out, and the client picked up the fountain pen . . . and it happened! The pen failed to work after two, then three attempts. Worse yet, on the fourth try a flood of ink poured out of the pen and over the insurance contract!

Lewis Edson Waterman hurried back to his office as fast as he could, to prepare a new contract form. But it was all in vain. When he returned, the client had already signed with another agent. This misfortune left Waterman no peace. Along with his brother, he looked for ways to improve the pen, for a new design principle that could automatically deliver an even flow of ink from the reservoir, without either drying up or spattering. Waterman hit upon the right idea very quickly. He had to make sure that the ink could always flow out of the reservoir in an even amount and, at the same time, air could simultaneously flow in from the pen to the ink reservoir. And that could be accomplished most effectively by using the capillary principle.

He was able to bore additional holes, as thin as hairs, in the ink conductor, the piece between the pen and the reservoir. The result: when the pen was placed on paper, ink was conducted to the writing surface. According to the need for ink, the natural surface tension of the ink allowed only as many tiny drops of ink as were actually needed run to the slit in the pen's nib from the capillary channels. At the same time, small air bubbles could rise into the ink reservoir via the channels and prevent a vacuum. When one stopped writing, the surface tension kept the ink from leaving the chamber.

The principle worked faultlessly. The ink flowed smoothly from the pen as if the point had just been dipped into an inkwell. There was no stopping, neither too much nor too little ink. The first really usable fountain pen was born, and was patented in 1884. Following the old rule of all constructors, "the fewer the parts, the fewer the problems," the "Waterman Ideal Fountain Pen" consisted of only four parts: the barrel, which simultaneously served as the ink reservoir (A), the front section (B), the ink feed (C) with its capillary channels that began with three branches and ended with two, and the pen point (P) itself, which fit between the ink feed and the front section and was supplied with ink according to the principles of capillary action and the force of gravity. All the parts were made of hard rubber, which could easily be shaped, bored, smoothed and polished. Filling was very

Christmas advertisement for the insertion pen, 1907.

simple. One unscrewed the front section and filled the shaft with the help of an eyedropper, a process that was to stay in use for decades.

Lewis Edson Waterman's invention caused a sensation. From the corner of a New York cigar kiosk at the entrance of the office building at 136 Fulton Street, where the inventor personally sold his weekly production of 36 pens on the weekend, there soon developed a production facility for which a six-story building on Broadway was scarcely big enough. The L.E. Waterman Co. became the world's greatest manufacturer of fountain pens and remained so until well into the 1930s.

Waterman's thoroughly successful technical approach soon gained scientific verification as well. James P. Maginnis reported in *Scientific American Supplement* on June 14, 1906, on the way a fountain pen worked from the standpoint of physics:

"It seems to be a very simple thing to construct a pen that holds a large supply of ink, especially when it consists of a casing made of a glass tube, pointed at one end, of small diameter. The tube may be filled with ink to the point at which it will hold no more air. If the opening at the writing nib is narrow enough, the ink will be held in the tube as a result of its surface tension. When one moves the writing nib over the paper, a thin film of ink is drawn from the opening and will form lines, letters, words, and sentences . . . Now many will think they have a fountain pen that they can carry with them in their pocket and use whenever they want. The laws of nature

Steps in manufacturing a Waterman "Clip-Cap," 1907.

will contradict them. It is more than likely that the possessor, the first time he takes his pen out of his pocket to show to his envious friends, will have to realize that the pen has emptied completely. Perhaps he thinks that a suitable closing, such as a piece of cork stuck on the open end, would prevent this misfortune. But only up to a certain point, for the greater warmth of the human body in comparison to that of the ink would make the ink in the tube expand and run out through the writing opening . . .

"The ink feed is for the most part responsible for neat, quick writing. When a filled pen is held with the point downward, a number of forces affect the ink inside it, but mainly the forces of gravity, inertia, capillary attraction, air pressure, friction, and the viscosity of the liquid, as well as other powers. A well-made fountain pen keeps these forces in equilibrium, and the ink does not run out. As soon as the pen touches the surface of the paper, the capillary forces change—the ink flows, the pen writes."

In his report, Maginnis raises the question of the ink's equilibrium, which could cause the pen's owner major problems, especially in cases of sudden temperature or air-pressure changes. When a pen is carried in the jacket pocket or held in the hand, the body temperature, which is at least ten to fifteen degrees Celsius greater than the room temperature, is transmitted to the pen. As a result, the air in the ink reservoir expands—all the more so if there is more air in the holder as a result of long writing or incomplete filling. The resulting higher pressure pushes the ink through the ink conductor to the pen point and drops form. A similar situation arises with major changes in air pressure, during a trip into the mountains, for example, or a sudden change between high-pressure good weather and low-pressure stormy weather.

The English fountain pen manufacturer Mabie, Todd & Co. improved Waterman's fountain pen in a practical way that is still used today. As early as 1910, its "ladderfeed" had numerous equilibrium-equalizing chambers, which could take up excess ink after pressure changes and slowly give it off to the pen point without spattering.

Left: Jules Fagard took over the license to produce Waterman pens in France.

Right: The famous "Hundred Year Pen" of 1939.

The futuristic "C/F" of 1953. Designed by a car designer and filled with ink cartridges.

Waterman

Early self-filler, 1911.

Water advertisement, 1913.

It was in New York in 1883 that the former book salesman, shorthand teacher, and insurance agent, Lewis Edson Waterman, began a new chapter in the history of the fountain pen. Equipped only with a pocket knife, a saw, and a file, he "carved" the first fountain pen from the hickory wood of a broken wheel spoke. Its ink conduction system utilized the capillary powers of nature. Pens that could be filled with ink had been known for decades. They supplied the pen point, via one or two fine channels, with ink from the reservoir unreliably, usually too much or too little. They lacked an effective regulatory system. Waterman was able to create this system of equalization in a simple manner by cutting two or three grooves into his ink conductor in addition to the main ink channel. Thus it became possible for air to circulate to the ink reservoir. Now enough air could enter to replace the ink that was used by the pen. The capillary forces held the ink molecules together like a chain. Convinced by his discovery, which he patented in 1884, and encouraged by friends and acquaintances, whose pens he had promised to repair at no cost should they need it, he ventured to advertise for customers: "Waterman's Ideal Fountain Pen Regular—Five-Year Guarantee!"

Though there were only two hundred pens made in the first year, production of the handmade fountain pens rose to five hundred in the following year! In 1893 the shy, gentle, and not at all business-like Waterman was talked into an advertisement for the sum of $63.50. The demand promptly rose: The newly founded L. E. Waterman Company sold five thousand pens in 1888. In 1901, when the founder of the company died, a thousand Ideal Pens a day left the Waterman factories in New York, Newark, Seymour and Montreal.

Business contracts for Europe began with a sensational event. At the 1900 Paris World's Fair a Waterman fountain pen won the grand prize for excellence of quality. Waterman's nephew, Frank D. Waterman, visited the most important European countries in the first years of the twentieth century and in 1905, the same year in which Waterman produced the first fountain pen with a riveted-on clip, made an exclusive contract with the Londoner, Lawrence G. Sloan, who received the distribution rights for all of Europe. In 1907 Waterman came up with the idea of a safety fountain pen, which made it possible to transport the pen safely in any position. With that, the fountain pen was no longer a luxury article for business people or office workers, but became an everyday utensil for everyone. And the Waterman ranked as the uncontested best-seller. Whoever had a good opinion of himself purchased the black writing instrument richly decorated with gold and silver filigree.

Lloyd George, the British Prime Minister, signed the Treaty of Versailles on June 28, 1919 with a solid gold Waterman. The King of the Belgians, the Queen of Rumania, and even the Emperor of China were equally proud possessors of a Waterman, as was Charles Lindbergh, who carried an avant-garde Waterman in his luggage during his famous Atlantic flight. For a long time now, the simple insertion fillers had been superseded by modern self-fillers, and an incalculable variety of colors and patterns was on the market.

In Europe, the Frenchman Jules Fagard, after having sold Waterman fountain pens and mechanical pencils successfully in his homeland, as well as in Switzerland and Belgium, for twelve years, founded the Jif-Waterman writing instrument factory in 1926. The pens, made under license, soon dominated the French market. In 1927 Waterman of France first introduced a mass-produced ink cartridge, a slim glass tube, closed with a cork stopper. Its production began in 1935 in a modern ink factory in Issy-les-Moulineaux, not far from Paris.

After Jules Farard's death, his wife took over the business and led it, under the brand name of Jif (the initials of Jules Isidore Fagard), to a period of economic success.

At the end of the 1920s, the Art Deco style influenced the design of writing instruments. Never before had more beautiful writing instruments been manufactured. The Waterman *Patrician* and *Lady Patricia* are still regarded by connoisseurs today as examples of perfection and classic elegance. The rest of the line bore numbers like those of Coco Chanel's famous perfumes: "Number 5," "Number 7" or "Number 92." Gold and silver caps and shaft rings were decorated with engraved flowers and leaves. Some customers are said to have ordered pens that were set with sparkling diamonds. Waterman prepared cases of pure silk for the ladies' models.

Along with the assortment of pens, Waterman stressed the further development of technology. In 1935 Jif-Waterman was one of the first firms to succeed in producing a usable ballpoint pen.

In 1939 Waterman introduced an unusual series of pens in America, the *100-Year Pen*. Its streamlined contours, the ribs running around it, and its glowing material of synthetic glass quickly made it a public favorite.

During World War II the fortunes of the traditional brand sank. Management errors, an unfortunate model policy, and pressure from the ballpoint pen that became popular in 1945 brought Waterman of America and England to the brink of collapse. Jif-Waterman in France, though, succeeded in beginning ballpoint pen production at the right time. The *Pantabille* refillable multicolored ballpoint pen, introduced in 1947, ranked among the first European successes of the new writing instrument generation.

In 1953 the "CF" was born. A General Motors automobile designer gave the instrument the look of a futuristic spacecraft. The most remarkable feature for the chronicler: The "CF" was to pave the way for modern filling with plastic ink cartridges.

In 1958 the company, which had made fountain pen history for 74 years, closed its doors. The brand-name rights for North and South America were sold to Baron Marcel Bich of Paris for one million dollars. Jif-Waterman, now under a new proprietor in Paris, tried to keep the company afloat in France and several other European countries. After Madame Fagard's death in 1964, her daughter, Elsa Le Foyer, took over the direction of the firm. Soon it looked as if the surviving Waterman firm would also have to give up.

Then in 1969 Madame Le Foyer's daughter, Francine Gomez, decided to take over Jif-Waterman. And she reorganized the company from the ground up. The management was rejuvenated radically. An assortment of new models for school children and an impressive one for adults were developed and put on the market with great success. In 1971 Jif-Waterman bought the brand name back from Baron Bich and changed the firm's name to Waterman S.A. Models such as the *Graduate* (1971), *Gentleman* (1974), *Jif 22* (1980), and *Man 100* (1983), the last honoring the hundredth anniversary of the Waterman brand, made the Waterman firm the unchallenged leader of the market in France and a worldwide specialist in high-quality, elegant writing instruments. In 1987 the Gillette Group purchased the majority of the firm's stock.

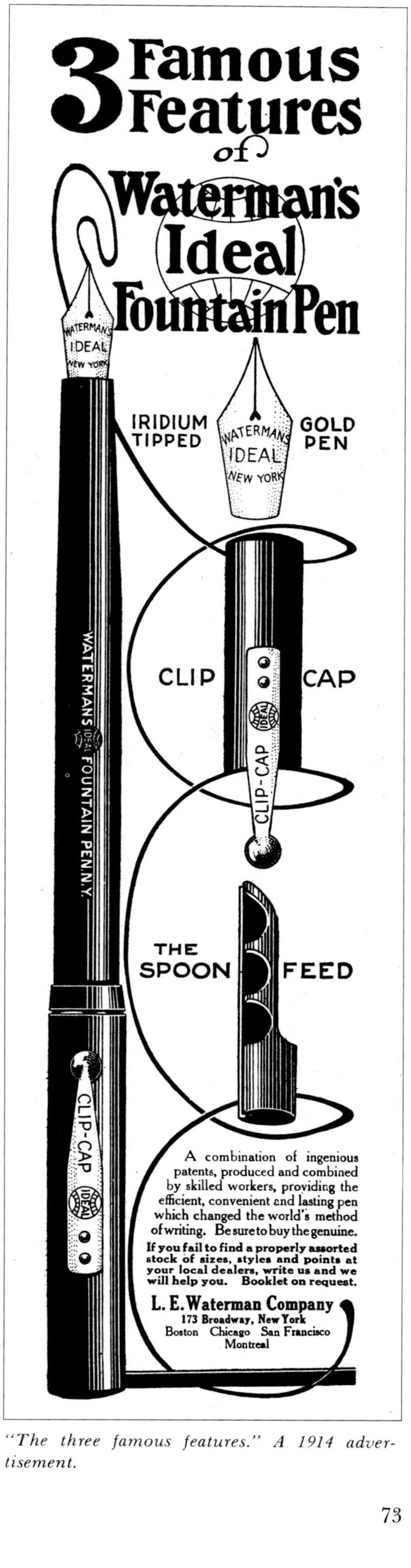

"The three famous features." A 1914 advertisement.

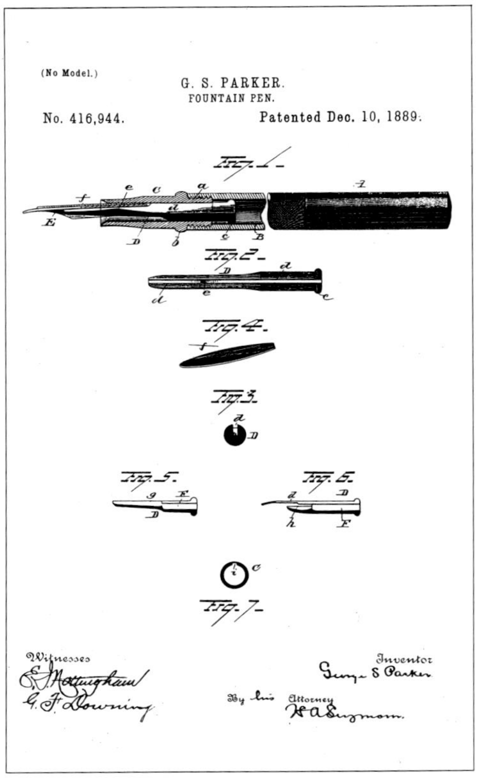

Paying a fee of five dollars, George S. Parker acquired the patent on his first fountain pen on December 10, 1889.

Lewis Edson Waterman's "Ideal Regular Fountain Pen" was the best-selling filler pen at the beginning of the twentieth century. Imitations popped up out of the ground like mushrooms. Dozens of already established or newly founded fountain pen manufacturers tried to market their concepts of more or-less-perfect fountain pens.

Paul E. Wirth, Lincoln, Laughlin, de Witt, Meredith, Keystone and Thomas A. Edison were among the best-known American producers who tried to cash in on L. E. Waterman's success. In Germany, Soennecken remained the most important manufacturer of fountain pens; Schwan and Roeder limited themselves to importing American products. In England, brands like Onoto and Swan won fame that stretched to the Far East. All of the manufacturers competed by offering special refinements, engravings, and decorations on the outside of the black hard-rubber shafts and caps. There were black, spotted brown, cylindrical, six- and eight-sided pens, plus those with gold bands. Others were thick in the middle and pointed at the ends.

At the beginning of our century, a standard fountain pen cost a dollar in America, about ten Reichsmark in Germany. The customer had to lay out somewhat more if he valued chased ornamentation and gold trim.

Among the already established American manufacturers there was already the John Holland Gold Pen Company, founded in Cincinnati, Ohio in 1841. John Holland pens, with their patented elastic points, were the preferred writing instruments in schools and universities.

George S. Parker—From Teacher to Inventor

In the 1880s, the sale of fountain pens was one of the more welcome possible ways to increase a teacher's somewhat meager salary. The young George S. Parker, a trainer at a telegraph school, had specialized in the John Holland brand. His sales to his numerous students went well. That practical instrument, the fountain pen, fascinated the students on the one hand and disappointed them on the other. Again and again, students took leaking or malfunctioning "Hollands" to their teacher and dealer and asked for help. Requests of this kind were part of a day's work. The buyer of a fountain pen in the 1880s accepted it as the price of the advantage of not having to dip the pen in the inkwell.

An advertisement for the American Caw brand, which was sold by the Schwan Pencil Factory of Nürnberg, 1904.

The Breakthrough

George S. Parker regarded it as his natural duty to help his students and clients. Although he knew very little about the physical laws after which a fountain pen functioned, he took the reclaimed devices apart and looked for the causes of the problems. And soon Parker found the main cause of the ills. The ink passages had to be improved. Parker got himself a few simple tools, a workbench, a thread cutter and the like, and put his students' fountain pens back into service by making small improvements. The more he experimented, the clearer it became to him that the ink flow of his "repaired" pens was more even, but they still did not achieve trouble-free writing. His gradually-gained knowledge only resulted in a perfect pen when he built one from the ground up.

Parker made up his mind. He cut, sawed, and filed a new kind of ink passage, mounted it in a barrel, and was happy to find that it worked far better than anything he had previously achieved by making repairs.

Paying a five-dollar fee, George S. Parker patented his first fountain pen on December 10, 1889. The telegraph teacher began to spend his evenings producing his own fountain pens in his small hotel room.

The early Parker products, like many other fountain pens, had their ink passages both above and below the pen's nib. Parker was one of the first manufacturers to decide, in 1891, in favor of the now-customary design, with the ink feed under the nib. And he was successful with it: The nib, now fully visible, gave the pen more elegance. The decorative value of the sparkling gold nib increased the fountain pen's value to the consumer.

Numerous improvements followed. In 1894 Parker invented a new ink feeding system, whose ends ran out in a bow and touched the wall in the interior of the ink reservoir. He called his invention "Lucky Curve." It is still known by connoisseurs today as one of the most famous functioning parts in the history of the fountain pen.

Its advantage was that the capillary forces and attractions that come into being at the place where the ink passage touches the inner surface of the ink reservoir pull the ink into the reservoir as soon as the pen is carried upright in the pocket. The ink is thus prevented from running out through the passage and thus making a mess of the cap.

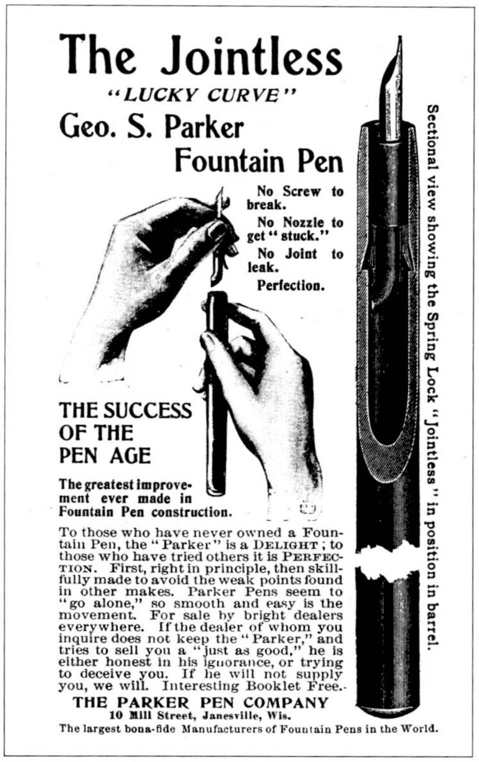

The famous Parker "Lucky Curve" ink passage. It prevented ink from coming out of the closed pen. An 1898 advertisement.

Above: The big American firm of Mabie, Todd & Co. advertised under the Swan brand name in 1905.

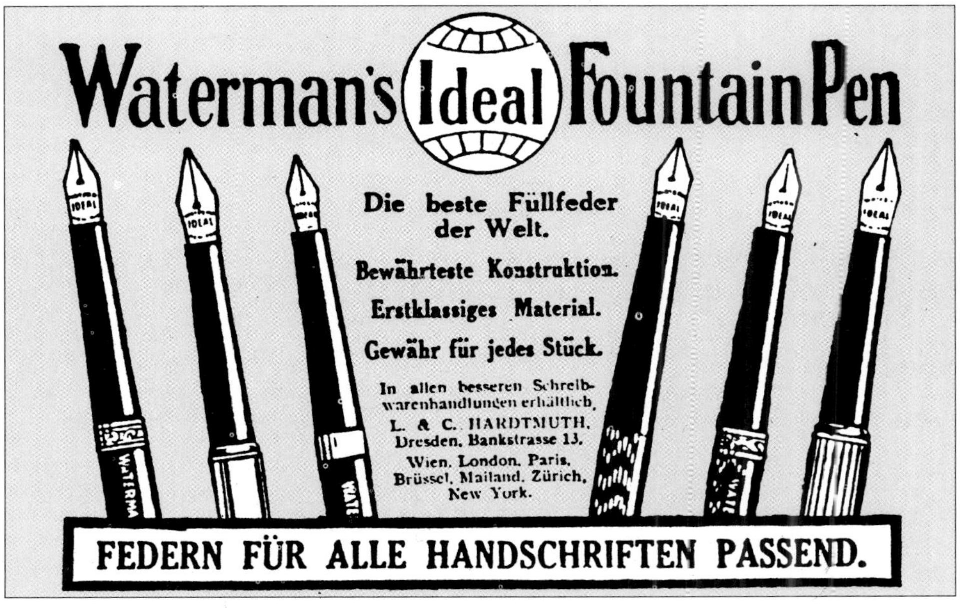

Left: Waterman's Ideal Fountain Pen became the leading international brand of fountain pen.

Parker lengthened the cap, which now reached over the front of the pen as far as the barrel. In 1899 Parker advertised in numerous newspaper advertisements for his first fountain pen with an unthreaded front section. This new type of snap closing eliminated the weak point of most fountain pens of the time: Ink often leaked out through the threads that connected the grip section with the barrel, staining the writer's fingers.

Around the turn of the century, one filled a fountain pen by unscrewing the forward section, including the point and ink conductor, and dropped the ink into the barrel with an eyedropper. This laborious and messy procedure was eliminated by the inventive Parker in 1904 with a return to the filling principle of the Automatic and Conklin. He set a long, thin rubber tube into the interior of the shaft, its opening surrounded by the end of the ink feed. When one pressed a round button projecting from the barrel, a long metal rod pressed the rubber tube and pushed the air out. Now one needed only to dip the pen into an ink bottle and slowly release the button that came out of the barrel again. As a result of the low air pressure inside, the ink pushed through the ink passage into the rubber tube and filled it.

The "Self-Filler" caused a sensation. Even the Salvation Army could not keep out of it. In its organization it offered the "Post Pen," a reliable device whose filling principle resembled the sucking principle of the air pump. To be sure, the officers and soldiers of the Salvation Army did not know how to operate the pen or repair it. Besides that, they probably did not possess the right business sense to market their Post Pen. In short, the Salvation Army fountain pen, legendary today, soon disappeared from the market very abruptly. Its filling system proved itself in the pens of other manufacturers. The English firm of Thomas de la Rue used this "spray filling" for the successful Onoto fountain pen. L. E. Waterman was concentrating more intensively on the commercial development of his firm than on the technical development of his products at this time.

How It Became a Tool

The fountain pen was no longer limited to a few business people, students, and professional writers. It became an everyday tool for everybody.

Waterman's factories were opened in numerous American and Canadian cities. Business connections were made with England, France, Belgium, Germany, Austria, Switzerland and Scandinavia. The success of the regular-type pen made Waterman hesitant about bringing out a "self-filler" too. The well-known manufacturer of *The Conklin Pen*, from Toledo, Ohio, had been promising "filling by a press of the thumb" for years. One pushed a heavy semicircular brass ring into the shaft, which in turn pressed a rubber tube in the barrel. After being released, the tube filled nicely with ink. A wide semicircle of hard rubber that could be moved sideways served as a barrier and prevented the unintentional movement of the half-moon filler ring. Many might have found this unusual device disturbing—others swore by their "crescent fillers," which ranked among the most popular American pens until the 1920s.

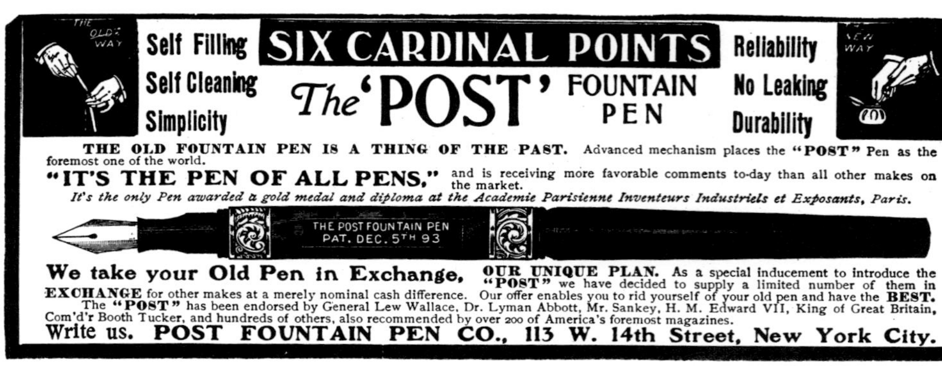

The "Post" pen of 1902. The Salvation Army handled sales.

L.E. Waterman observed the many ideas and attempts of his competitors to provide problem-free refilling with ink, and felt justifiable mistrust. His own wish to gain success in the market in 1903 with a piston pump in a transparent barrel had been a failure and soon had to be abandoned. Above all, presumably, because it was not possible to make the piston slide up and down along the inner wall of the ink reservoir without leaking.

Clip-on pens like Waterman's regular pen suffered from another "teething trouble:" Although they were always to be carried with the cap upward, it happened now and then that ink ran out accidentally and made a mess of the interior of the cap, the front section, and ultimately the owner's clothing.

A new system, based on an 1898 invention by Brown, eliminated this problem. The "safety filler" was completely different from all other fountain pens in use at that time. The *Safety Pen* was the first fountain pen to guarantee protection from leaking ink, no matter in what position it was carried or what other influences, such as a change in the temperature around it, might occur.

These important advantages helped the safety fountain pen become one of most popular types of pen in America and Europe. So it is no surprise that its functioning principle was gradually adopted by most manufacturers, even though one major disadvantage remained. The safety pen did indeed offer protection from messy leakage, but filling it with an eyedropper after unscrewing the pen's nib remained, as before, a laborious, time consuming process, accompanied by the danger of spilling ink.

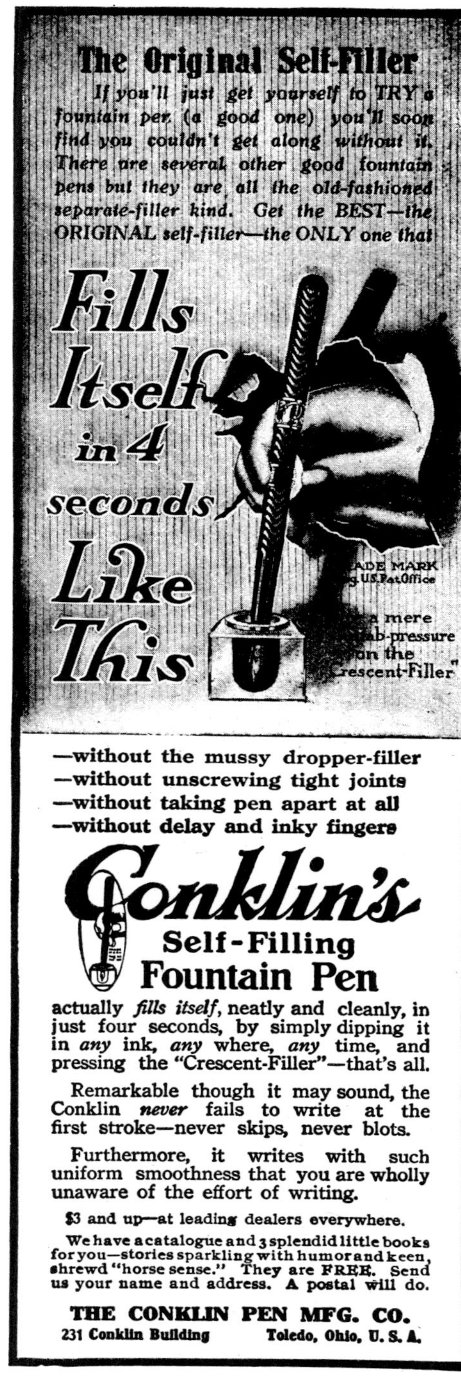

The father of all self-fillers: the Conklin with the metal bar in its barrel.

First appearance in Germany: Parker's introductory advertisement of 1928.

Parker

"Fountain pens can write in any language" was the motto of George Safford Parker, the inventor of the Parker fountain pen, and with this motto, Parker wrote fountain pen history in the truest sense of the word. His firm, now ranking among the leading international manufacturers of writing instruments, with factories in North America, Great Britain, and France, had its origin in a small hotel room in Janesville, Wisconsin. Here the telegraphy teacher repaired his students' fountain pens, and here the idea was born of making a fountain pen with an improved ink passage by himself, by drilling, filing and sawing. His success encouraged him to found his own company in 1888, a century ago. One year later his first patent was granted. Hundreds were to follow.

George S. Parker was a tireless experimenter. With almost every succeeding year he was able to make improvements to the ink flow, the screw connections, the filling apparatus, the cap, or the ink passage. The *Lucky Curve* model, patented in 1894 as a revolutionary ink passage, was produced almost unchanged for more than thirty years. Among the most famous owners of Parker pens were the Italian composer Giacomo Puccini, who wrote the opera "La Boheme" with one in 1896. Sir Arthur Conan Doyle used a Parker pen when he created the famous detective Sherlock Holmes, and when he wrote dozens of detective stories toward the end of the nineteenth century. As a telegraphy teacher, Parker was well acquainted with the basic human need to communicate with others. His market, he decided, should be every person in the world who wanted to express his thoughts clearly, easily, and enduringly.

In 1918 the Parker Company took in one million dollars for the first time, a fraction of what it spends on international advertising today. In 1919 a modern five-story factory building was built to house the firm's constantly growing production. To the astonishment of the whole pen industry, Parker had the courage to put the *Duofold* on the market in 1921, a pen that cost more than twice as much as a normal fountain pen, all of seven dollars. In addition, the *Duofold* brought color into the industry which had made practically nothing but black pens until then. Its bright orange-red color was so unique that Parker was one of the first firms to patent something other than the product. Parker patented the color!

At a time when only about five percent of those Americans who could read and write owned a fountain pen, the new Parker *Duofold* inspired enthusiasm. Its popularity and allure proved to be so lasting that forty years later a ballpoint pen known as the "Big Red" could be sold by the millions with great success as a throwback to the "Roaring Twenties." Around 1924 the Parker export organization felt strong enough to put out feelers toward Europe. In 1925, along with the Osmia firm, a sales branch was founded in a German writing-instrument center, at Dossenheim near Heidelberg. In 1928 Parker put the improved *Duofold* in glowing colors, made of unbreakable plastic, on the German market:
"The famous fountain pen has just arrived in Germany." But the new colors led to some surprises on the international market. In India green pens were a failure, because green is regarded there as a symbol of misfortune. White pens were rejected in China and Japan, because white is the color of mourning in those cultures.

Like most self-fillers, the Parker *Duofold* also had a tube as its ink reservoir. It was compressed by pressing a button and then released to fill with ink. This popular but not very reliable method was abolished in 1933 with the introduction of the Parker *Vacumatic*, the first tubeless Parker fountain pen. The *Vacumatic* could take in 102% more ink than the *Duofold*, and charmed buyers with its design, made by a famous New York artist. The Parker *Vacumatic* was the first model to bear the form of an arrow on its nib and as a clip, which later became a Parker hallmark.

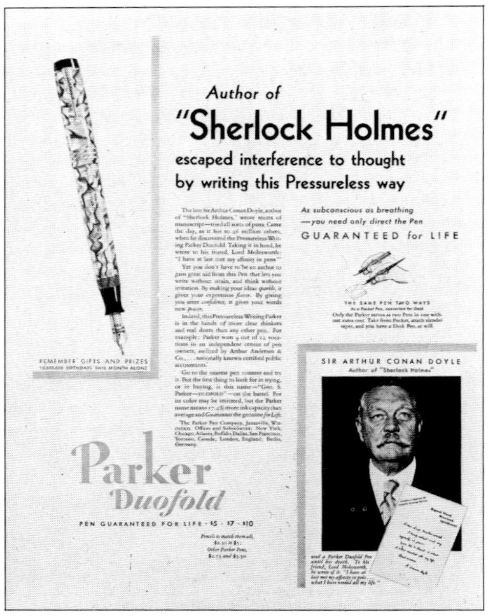

Famous authors wrote their works with a fountain pen; 1921.

In 1940, as Parker was about to introduce its next major development, it was decided not to use an additional name that drew attention away from the name of Parker and could be translated into other languages only with difficulty, if at all. The product was called "Parker 51."

The unusual, futuristic design and the new materials made it one of the five best designed modern products in America, as a questionnaire determined. In another survey, *Time Magazine* found that Parker pens, Coca-Cola, Singer and the Jeep were the best-known symbols of America overseas.

During World War II the sale of Parker pens in Germany and elsewhere in Europe had to stop. After the war, the new ballpoint pen caused the decline of the fountain pen as a product for the masses. Parker refused to start producing it. The rolling ball that still made spots and smears seemed too underdeveloped and thus damaging for the market image. It was nine years after the first Biro and Reynolds ballpoint pens that the Parker *Jotter* appeared, with a large-size cartridge. Improved with a tungsten-carbide ball in 1957, the Parker *Jotter* became a worldwide best-seller, as it still is today.

A year before that, the Parker 61 fountain pen was introduced. Unlike its predecessors, it no longer had any moving parts, but filled automatically by capillary action. In the same year Parker took over the Eversharp Pen Company and made it into the low-price section of the firm.

The year of 1963 has a special place in the history of the Parker firm. The company celebrated its 75th anniversary. The famous Parker 75 was introduced, and a branch was opened again in Baden-Baden, as Germany remained one of the most important fountain pen markets in the world.

The introduction of the Parker 75 was accompanied by a unique advertising campaign. A limited edition of this pen was sheathed in silver from a treasure that had sunk with a Spanish fleet in a 1715 hurricane off the Florida coast and been salvaged by Kip Wagner 250 years later.

The German Parker Pen Company, along with its Luxor Pen branch, offered a significant assortment in 1981: Fancy writing instruments, like the pendant *Swinger* ballpoint pen and the *Blimpy* stand-up pen became successful internationally. From England came the unique stainless-steel Parker 25 models.

Meanwhile, the Parker Company had grown into a firm with over 6000 employees. Along with writing instruments, the firm owned a temporary employee service (Manpower) and a company that producedleisure time games.

The Parker 75 was followed in 1967 by the Parker Classic line, in 1981 by the Parker *Arrow*, and in 1983 by the *Premier*. In 1986 the owners of the Parker firm separated from their writing-instrument division. Parker Pens became the property of some of the firm's former management personnel. The formerly American firm moved its headquarters to Newhaven, England.

For its hundredth anniversary in 1988, Parker offered its loyal customers another surprise. The *Duofold* of the 1920s enjoyed a comeback as the pen of the century, with modern technology inside, but as beautiful as ever on the outside.

The tubeless "Vacumatic" won millions of friends. A 1938 advertisement.

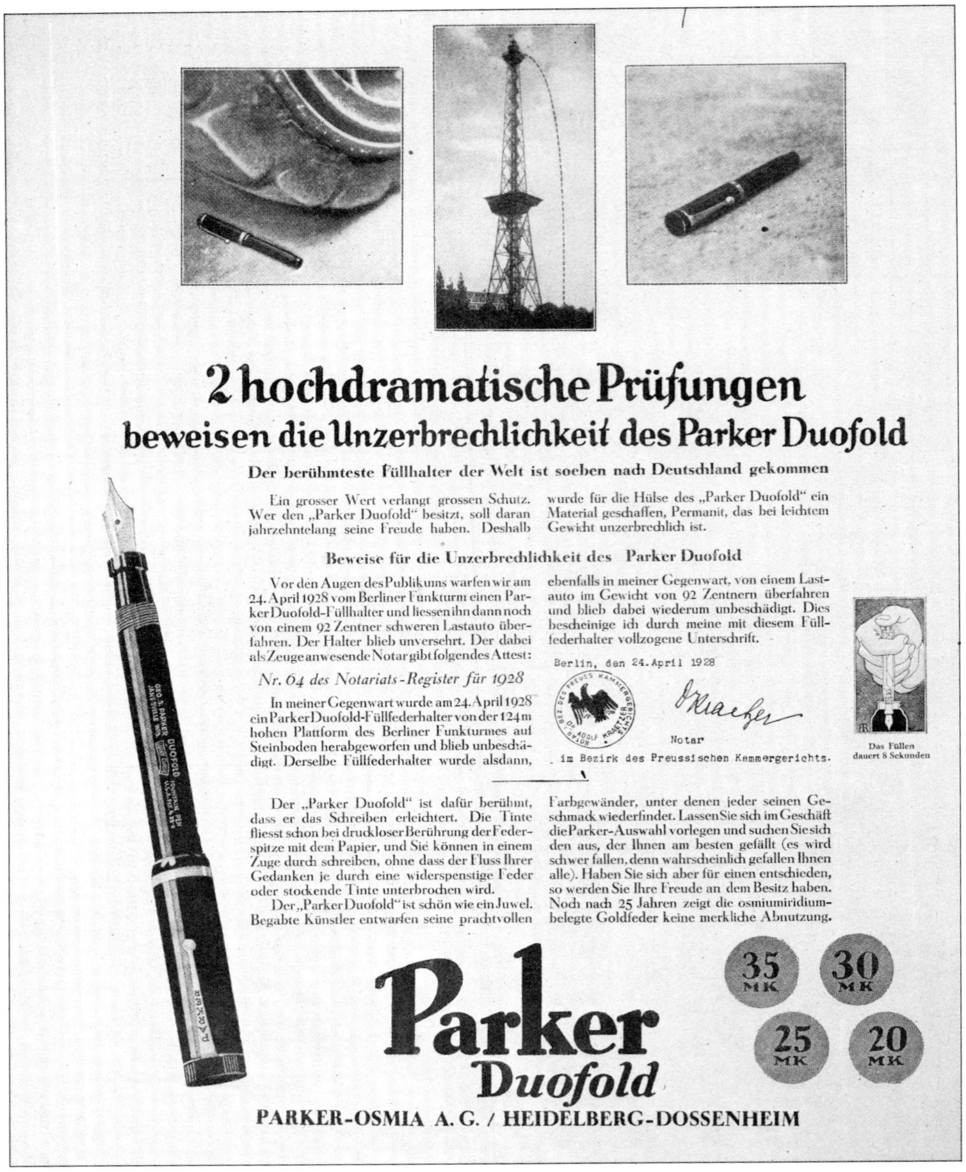

The unbreakability of the Parker "Duofold" was demonstrated to the German public impressively in 1928.

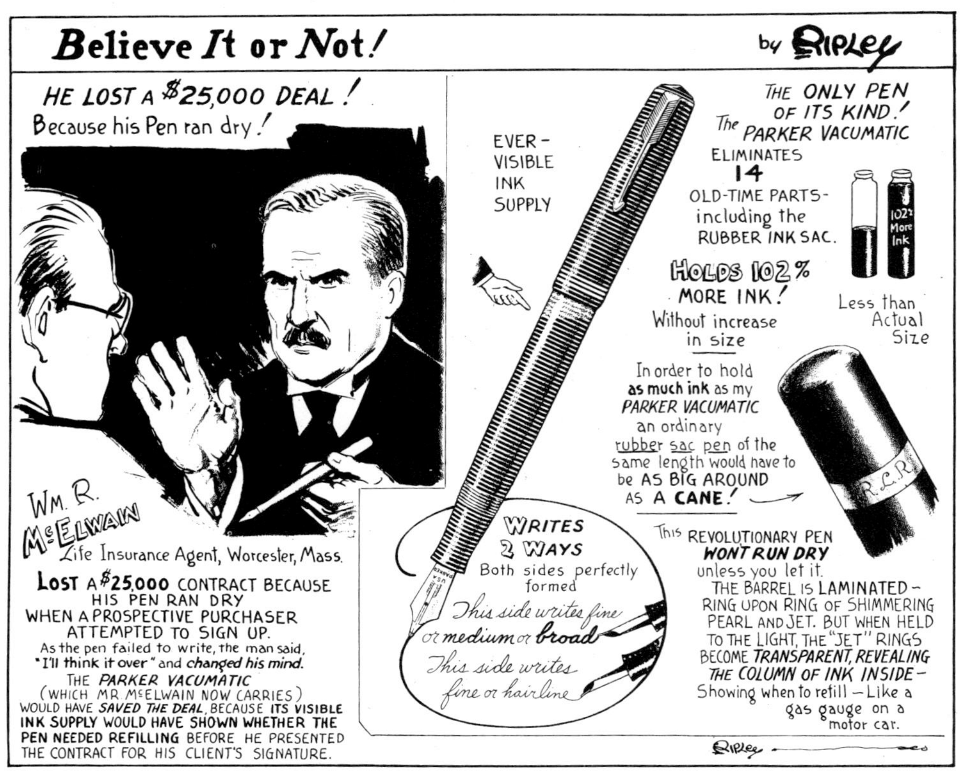

Unbelievably, the Parker "Vacumatic" held 102% more ink (1934).

Die abenteuerliche Story des Parker 75 Treasure Pen

1715 war es, als die spanische Armada, die berühmte Silberflotte, bestehend aus 11 mächtigen Galeonen, in einem fürchterlichen Hurrikan vor der Ostküste Floridas (dem heutigen Kap Kennedy) unterging.
Über 700 Matrosen fanden den Tod. Nur 1 Schiff konnte aus dieser Katastrophe entkommen. Die stolze Flotte König Philips V. war auf der alljährlichen Reise von den Westindischen Inseln (dem heutigen Kuba) nach Spanien und führte Schätze, Münzen, Gold- und Silberbarren, Seide und wertvolles Porzellan im Werte von über 50 Millionen DM mit sich.
Einen großen Teil des Schatzes konnten die Spanier zwar noch im gleichen Jahr bergen — auch Piraten von den Bahamas holten sich ihren Teil — doch blieben Schätze und Material von unglaublichem Wert auf dem Grund des Meeres. 250 Jahre lang.

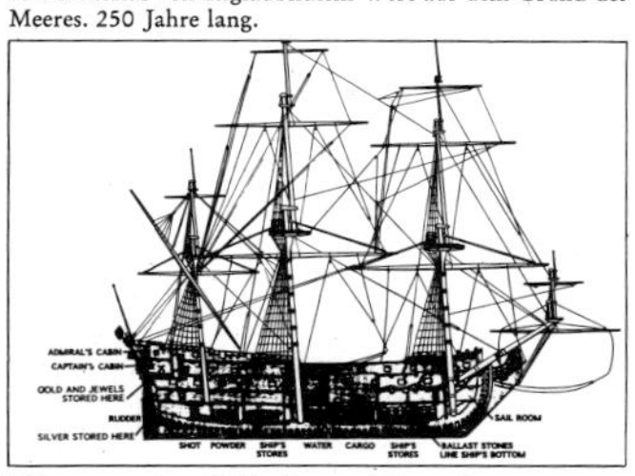

Ein unglaubliches Abenteuer beginnt

Kip Wagner, ein „verschrobener Antiquitätenhändler und Nichtstuer", wie er sich selber nennt, machte einen erstaunlichen Fund. Eines Tages schlenderte er am Ufer des Meeres nördlich von Vero Beach/Florida entlang, als er etwas Helles entdeckte. Es war ein Stück Silber, das mit einem Kreuz und einem spanischen Wappen gestempelt war. Eine Silbermünze. Ein Hurrikan, der einige Tage zuvor an dieser Stelle gewütet hatte, warf diese Münze ans Ufer.
Dieser Tag brachte eine dramatische Wende in sein Leben: Kip Wagner sollte der größte Schatzfinder seit 1687 werden.
Kip Wagner sammelte und studierte alles, was an historischen Niederschriften aus der Zeit des Unterganges der Silberflotte vorhanden war. Unter anderem auch die alte Karte, die ihm zeigte, wo „ . . . der Admiral, der die Flotte 1715 kommandierte, und der Rest der Schiffe . . . zwischen dieser Stelle und dem Bleech Yard versank . . ."

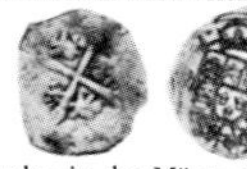

Diese Münzen wurden in der Münze von Mexico City hergestellt.

Er machte jetzt regelmäßige Besuche am „Geldstrand" und versuchte, mittels eines alten Minensuchgerätes den Boden gründlicher zu untersuchen. Mit Erfolg.
Er fand weitere Münzen, Schiffsnägel, Kanonenkugeln, Entermesser und — als besondere Krönung dieser Arbeit —: einen goldenen Ring, reich besetzt mit Diamanten.
Jetzt ergriff ihn das Jagdfieber. Er charterte ein Flugzeug und ein Spezialboot, das ein Sichtfenster nach unten hatte. Dabei entdeckte er alte Kanonen auf dem Meeresgrund, die mit einer dicken Kruste und Muscheln bedeckt waren. Nun galt es, eine Bergungs-Lizenz zu erwerben und eine gute Mannschaft zusammenzustellen. Es gelang ihm.
Der Staat Florida erlaubt dem Finder, 75% der gefundenen Wertsachen zu behalten. (Ein stolzer Anteil, wenn die Schätze lohnenswert sind.)
Mehr als 10 Jahre arbeitete die Mannschaft von Kip Wagner. Sie konstruierten Maschinen, die die riesigen Masse[n] Sand von den Wracks entfernten, tauchten tage-, wochen- und monatelang — so lange, wie es das Wetter erlaubte.

Jeder Tag brachte regelmäßig Funde. Mal weniger, mal mehr. Chinesisches K'ang-Hsi-Porzellan, über 250 Jahre alt (und vermutlich noch nie von einem Menschen benutzt), wurde — ganz unversehrt — geborgen, 1000 goldene Dukaten und viele Tausende Silbermünzen.

Kip Wagner schätzte seinen Fund auf mehr als 1 Million Dollar.
Doch lassen wir ihn seine Story zu Ende erzählen: „ . . . wenn ich auf die harten, arbeitsreichen Jahre zurückblicke, erscheint mir der Geldwert ziemlich bedeutungslos. Das Außergewöhnliche an unserer Geschichte liegt, glaube ich, darin, daß wir eine wirkliche Begegnung mit der Vergangenheit hatten . . . die Erregung der Suche, auch in den Monaten, wo wir warten mußten und schon halbwegs verzweifelt waren, haben Momente gebracht, die man mit Geld nicht kaufen kann. Jeder Fund erscheint als Geschenk aus dem Meer, und unsere größte Auszeichnung wird immer das unvergeßliche Erlebnis der Bergung sein."

Aus diesen spanischen Silbermünzen wurde ein selten schöner und wertvoller Füllhalter: Der Parker 75 Treasure Pen.

Parker Pen Comp., Janesville/USA, der Welt fortschrittlichster Füllhalter-Hersteller, kaufte einen Teil der Silbermünzen Kip Wagners.
Aus diesem 250 Jahre alten Schatz entstanden Parker-Füllhalter, die heute bereits einen ungewöhnlich hohen Sammelwert besitzen: Parker 75 Treasure Pens.
Auf dem jedem Füllhalter beiliegenden Certifikat bescheinigt Kip Wagner selber und ein Notar, daß es sich um das Original-Silber aus dem spanischen Schatz handelt, der 1715 vor der Küste Floridas versunken ist.

Der Unterzeichnete, aufgrund einer behördlichen Lizenz berechtigt zur Bergung von Schätzen und Kunstgegenständen, bestätigt hiermit, daß das in diesem Parker 75 verarbeitete Sterling Silber aus der spanischen Schatzflotte, die im Jahre 1715 sank, geborgen wurde.
Weiterhin wird bestätigt, daß dieses Silber das Zeichen der Münzanstalt von Mexico City trug und einen Teil des Reichtums darstellt, der von den Westindischen Inseln in die Schatzkammern des Königs von Spanien, Philip V., transportiert wurde. (. . . . . Kip Wagner)
Florida, District Indian River,
unterschrieben und vereidigt in meiner Gegenwart
am 10. September 1965 (. . . . . Notar)

Das „M" ist das Münzzeichen der Münze von Mexico City, der Adler das Zeichen Philips V.

Parker 75 - dieser kostbare Füllhalter jetzt auch in exquisitem Gold und strahlendem Vermeil.

Es ist verständlich, daß Parker nur eine begrenzte, unverkäufliche Anzahl 75 Treasure Pens anfertigen konnte. Das edle Silber war ja eines Tages aufgebraucht. (Heute ist der Erwerb eines Parker 75 Treasure Pens ein Wunschtraum eines jeden echten Sammlers.)
Aber eine ganze Reihe anderer Parker 75 — aus nicht weniger edlem Metall — können Sie heute erwerben. (Was den Treasure Pen von diesen unterscheidet, ist sein historischer, archäologischer Wert, der Hauch einer 250 Jahre alten Geschichte.)
Ein 14karätiger Goldmantel umschließt die Modelle des Parker 75 Insignia. Verlockend ist sein Anblick — repräsentativ sein Besitz.
Eine Goldverarbeitung ist jedoch bei vielen Parker-Schreibgeräten nichts ungewöhnliches — eher schon das neue Metall: Vermeil.
Es war nicht leicht, dieses Metall herzustellen, das im 17./18. Jahrhundert, zur Zeit des übergroßen Luxus, seine Entdeckung feierte. Vermeil ist eine Komposition von massivem Sterling Silber mit 14karätigem Gold. Diese außergewöhnliche Zusammenstellung bewahrt den warmen Ton antiker, kostbarer Metalle.
Ein Parker 75 aus Vermeil ist ein ungewöhnliches Schreibgerät. Nicht nur die Einmaligkeit seines Äußeren besticht, auch die Funktion dieses Füllhalters ist zukunftsweisend. Nehmen Sie ihn einmal zur Hand und prüfen Sie. Er ist es wahrhaftig wert. Er gehört einfach in jedes qualifizierte Fachgeschäft,
— und Sie sind doch eins!

Baden-Baden, im November 1966
Parker Pen GmbH

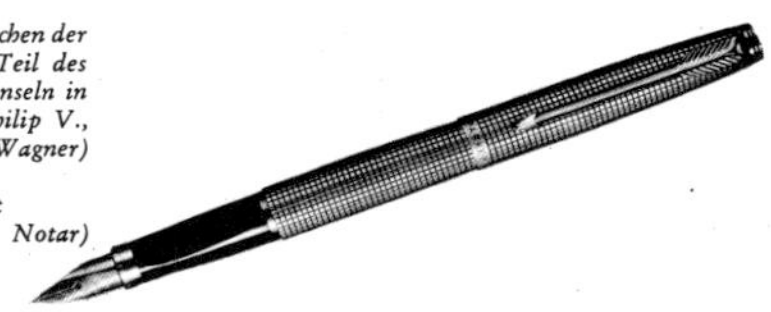

With a suspenseful story, Parker advertised its new Series 75 to specialty shops in 1966.

Safety filler of 1908: Montblanc-Simplo "Diplomat."

Protected by Patent

Not everything that is invented as new can be protected. And much that appears at first glance to be new turns out to be old stuff when examined more closely. "New," as the German patent law states, must apply to the idea, "a creative achievement" that has a "commercial application," if it is to be recognized by the testers at the German Patent Office in Munich. Within eighteen months of the application date,nton is published by the Patent Office, in case there are counterclaims that might make acceptance impossible. Only when third-party objections do not appear or can be rejected successfully in patent court does a patent become official and protects the invention from imitation for twenty years.

Patents for writing instruments from inside and outside Germany fill meters of files at the German Patent Office. Over 400 inventions from 1890 to 1940 apply to the fountain pen alone. Only a fraction of those that the official testers have recognized as worthy of protection are actually put to use. The examples shown here offer a far from representative overview of the patents of German and foreign inventors of writing instruments.

Original ideas were patented over the years but rarely went into production (left to right): Ink regulation by push-button—Fountain pen with two insertable nibs—Coil-spring ink flow—tube ballpoint pen—fountain pen with two horizontal nibs—fountain pen with two vertical nibs—ballpoint in a felt bed—cam-wheel ink flow.

The American Patent Office in 1851.

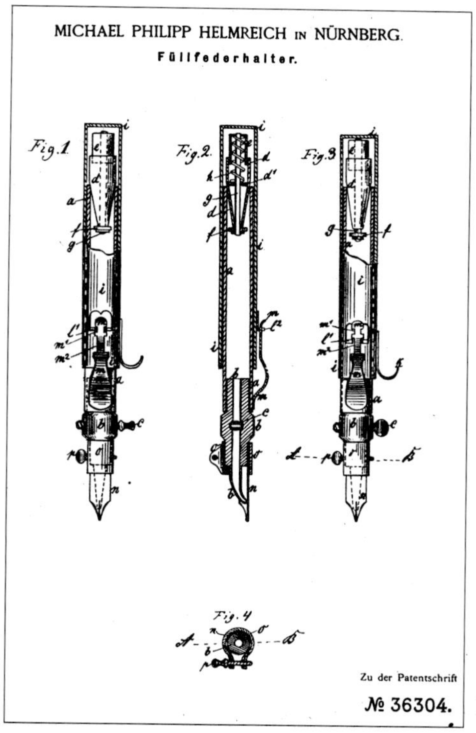
MICHAEL PHILIPP HELMREICH IN NÜRNBERG.
Füllfederhalter.
Zu der Patentschrift
№ 36304.

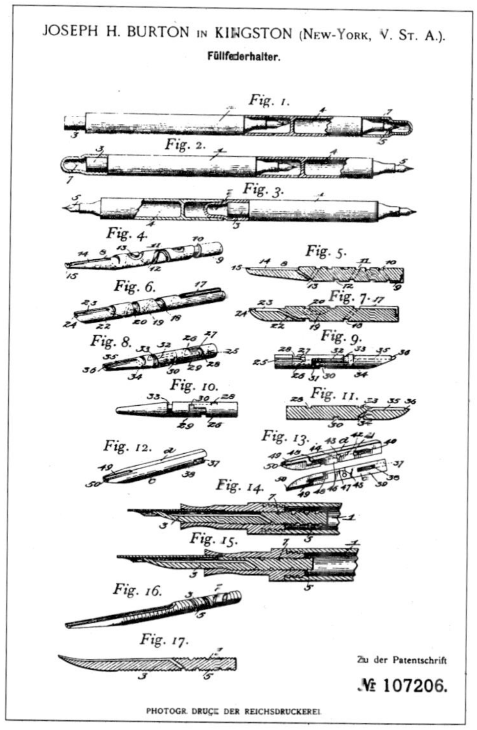
JOSEPH H. BURTON IN KINGSTON (NEW-YORK, V. ST. A.).
Füllfederhalter.
Zu der Patentschrift
№ 107206.
PHOTOGR. DRUCK DER REICHSDRUCKEREI

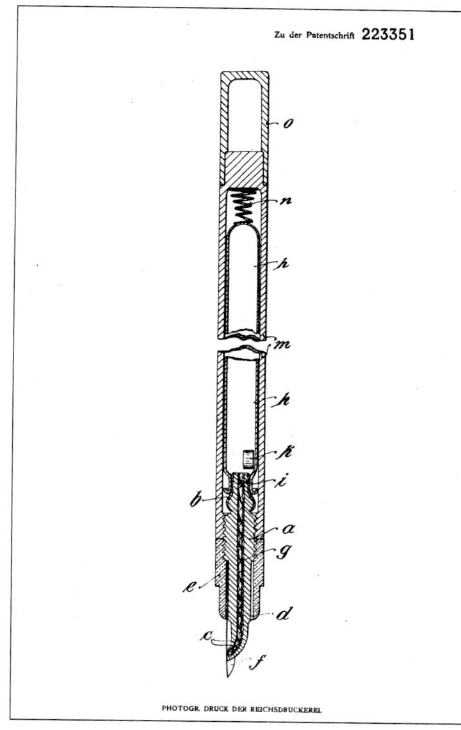
Zu der Patentschrift 223351
PHOTOGR. DRUCK DER REICHSDRUCKEREI

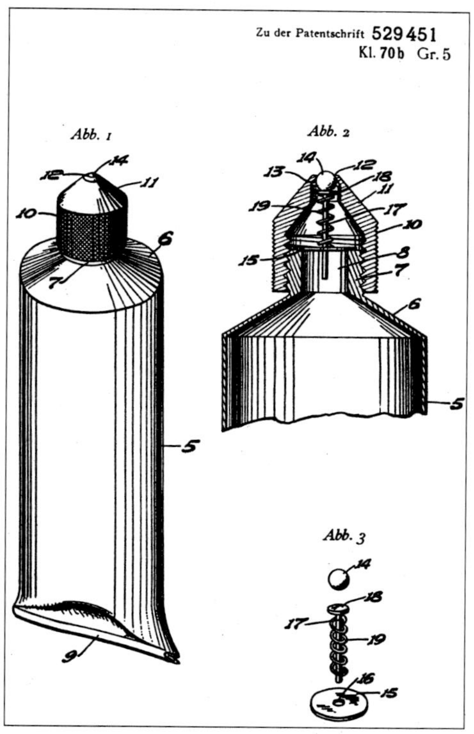
Zu der Patentschrift 529451
Kl. 70b Gr. 5
Abb. 1
Abb. 2
Abb. 3

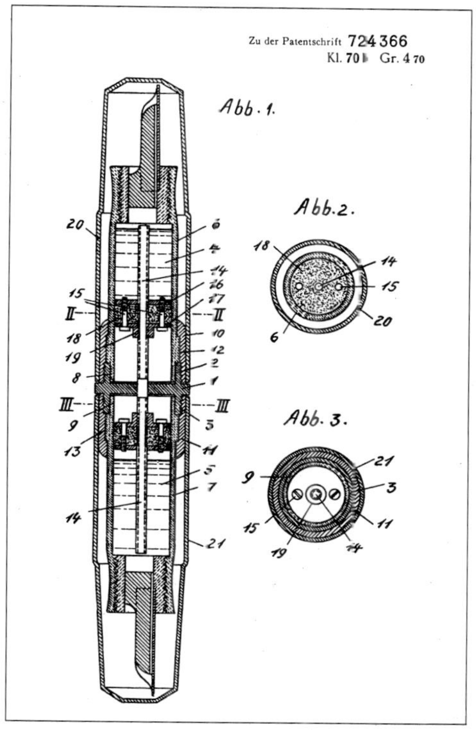
Zu der Patentschrift 724366
Kl. 70b Gr. 4 70
Abb. 1.
Abb. 2.
Abb. 3.

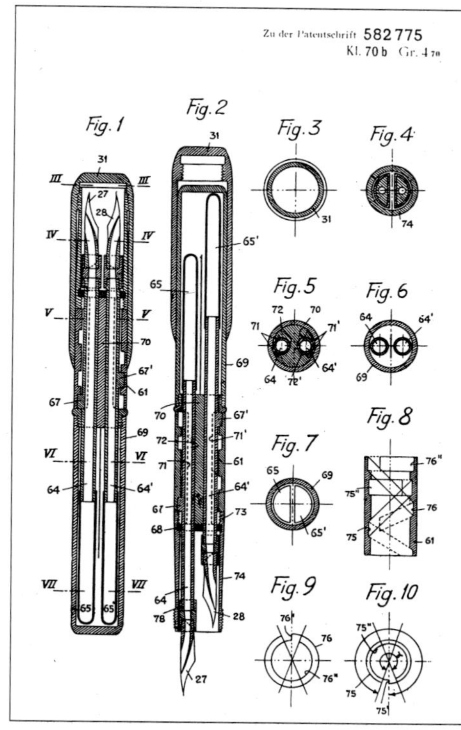
Zu der Patentschrift 582775
Kl. 70b Gr. 4 70

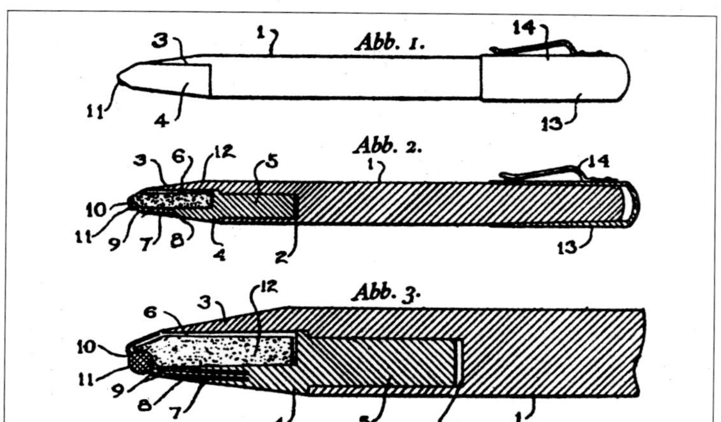
Abb. 1.
Abb. 2.
Abb. 3.

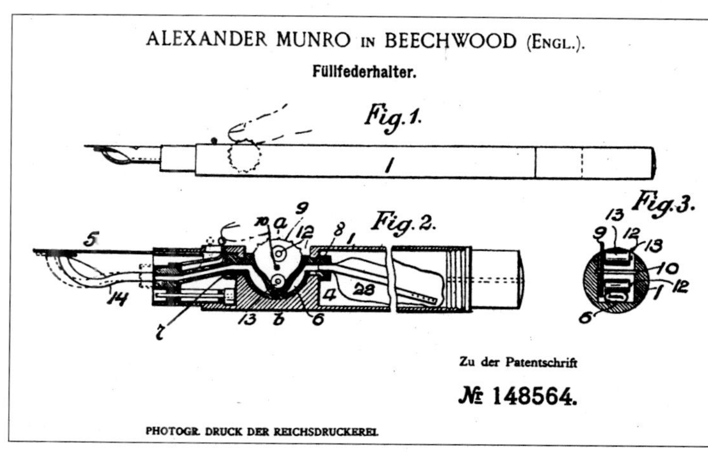
ALEXANDER MUNRO IN BEECHWOOD (ENGL.).
Füllfederhalter.
Fig. 1.
Fig. 2.
Fig. 3.
Zu der Patentschrift
№ 148564.
PHOTOGR. DRUCK DER REICHSDRUCKEREI

W. A. Sheaffer—The Clever Jeweler

The third outstanding personality among the American pioneers in the early years of the fountain pen likewise got acquainted with that "modern" writing instrument by chance. Walter A. Sheaffer already had much experience as a jeweler in his family's business, as a piano and organ dealer, and as an agent for a sewing-machine company. On an autumn evening in 1907, while looking through the local newspaper, he came upon an advertisement from the "Conklin Pen Company." In it, this firm from Toledo, Ohio praised their original filling method:
"Press a coin into the slit in the shaft—and the pen is filled in seconds!"

Sheaffer was fascinated by the idea of an "automatic" filling system instead of the eyedropper used to fill most fountain pens. But such a filling system really ought to be more elegant than the coin-operated mechanism that stuck far out of the barrel, as was offered in the advertisement. The very next morning he went to work in his small jewelry shop in Fort Madison, Iowa. He made a thin, flexible rubber sack out of soft tubing and set it in the barrel of a pen. The opening of the sack was narrow enough to be pushed over the end of the ink passage under pressure. Instead of the coin-shaped button, Sheaffer set a small lever into the barrel so that it did not project tangibly. When one folded the lever out with one's fingernail, a small metal bar pressed the rubber sack together. Now the pen just needed to be immersed in ink, the lever slowly released and pushed back into its groove in the barrel. Sheaffer's first fountain pen was clean and could be filled with one move of the hand. An amazingly simple principle!

Sheaffer was immediately convinced that his idea could give the development of the fountain pen an important impetus. On August 25, 1908 he received the patent for the first lever mechanism. But the cautious Walter A. Sheaffer knew how to control his own enthusiasm. The risk of investing his entire savings in the construction of a big fountain pen factory was too great. In view of the fact that almost sixty large and small manufacturers of fountain pens were competing bitterly with each other for control of the North American market, the danger of going bankrupt and, moreover, losing his prosperous jewelry business, was not worth the risk. To be safe, he made a handful of his new lever fountain pens and gave them to friends and acquaintances for testing. The results encouraged Sheaffer to invest capital from his savings and from his jewelry store profits, step by step, in building up his pen production. In 1912 he had reached the point at which he and two business partners began to manufacture lever fountain pens in Fort Madison. Seven employees crowded into the small workshop next to the jewelry store and produced the new Sheaffer fountain pens by hand. In 1917 there were already more than a hundred employees, who assembled over 100,000 fountain pens, going through 62 work processes per item. The gleaming lever on the barrel became the identifying mark of the leading Sheaffer "Self-Filler" of the early years. It was to remain so until well into the 1940s.

The success of the technically simple but particularly comfortable lever fountain pen compelled most of the other manufacturers to bring out lever-operated self-fillers too. In 1913 Waterman expanded its assortment of regular pens and safety pens with a "Self-Filling Lever Type." Others, such as Wahl-Eversharp, Carter, Conway Stewart, Onoto, Conklin, Swan, Sailor, Platinum, and Pilot, did the same. Only two of the larger fountain pen manufacturers stuck with their own systems: Parker in America and Pelikan in Europe. Parker, which had patented a mechanical filling system as early as 1904, improved its system at about the same time that Sheaffer introduced the lever system. Instead of the lever on the barrel, that might be seen as destroying the aesthetics, Parker offered a push-button at the end of the barrel. The filling principle worked similarly to that of the lever. By pushing the button, a flexible metal rod pushed against a rubber tube.

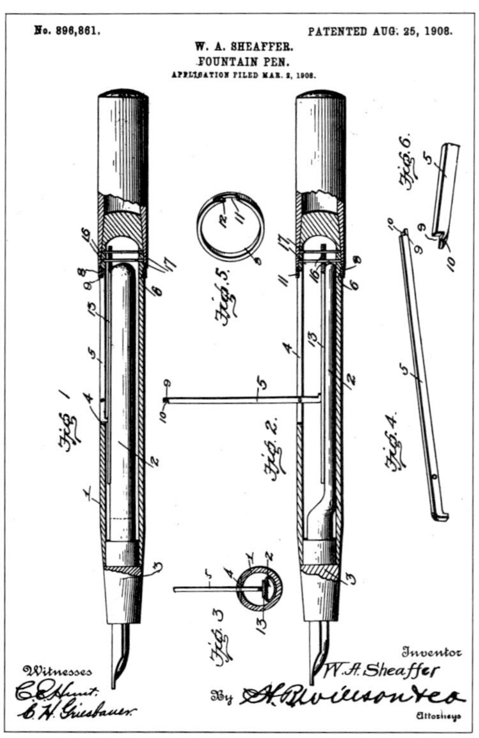

Sheaffer's patent for the first tube fountain pen with lever mechanism.

Before and during World War I, the use of the fountain pen enjoyed an unexpected increase. The American Expeditionary Corps fought alongside its allies in Europe, thousands of miles away from home. It was only right for every American family to remember "our boys over there" with presents. The fountain pen was an ideal gift, that also might encourage the boys to write home more often. The great American manufacturers realized this very quickly and competed for customers in patriotically tinged advertisements. L. E. Waterman and Parker (marketing the *Trenchpen*) also offered army and navy ink tablets that, when mixed with water, offered "Uncle Sam's fighting boys" the chance to keep in close touch with home anytime and anywhere without needing an ink bottle.

A welcome impetus to inspire mail from "our boys" or from home. A 1916 advertisement.

Sheaffer

Like most fountain pen pioneers, Walter A. Sheaffer, born in Bloomfield, Iowa in 1867, had learned and practiced several trades before he happened to get acquainted with the fountain pen and, so to speak, move its technical development forward by a giant step through a flash of inspiration. In 1907 Walter A. Sheaffer succeeded in creating a fountain pen where it was only necessary to move a small lever on the barrel up and down to fill the pen easily and quickly from an inkwell.

He had no way of knowing that the lever filling system he invented would be one of the most often copied and used filling methods of this century. And like most of the great names in fountain pen history, Sheaffer began very modestly. In the workshop of his small jewelry store and with the help of seven employees and Sheaffer's son Craig, the first handmade self-filling pens were produced. Every dollar that was earned by selling them was invested by the ambitious but cautious Sheaffer in the expansion and improvement of tools, equipment, and, later, small machines. In 1913 it had developed to the point where Walter A. Sheaffer could sell his jewelry store. The Walter A. Sheaffer Pen Company was founded in Fort Madison, Iowa. The new filling system quickly established itself. Almost all the established competitors felt compelled to copy Sheaffer's idea and circumvent his patent.

The year of 1924 marks another high point in the history of the Sheaffer firm. After a series of failures in 1920, Sheaffer used a new, unbreakable celluloid material, pyroxylene, that could be produced in any desired color and shape to make his pens. This innovation and Sheaffer's idea of using a white dot on the cap to symbolize a lifetime guarantee, soon brought the Sheaffer firm to the top among American manufacturers of writing instruments.

Five years later Sheaffer created the *Balance Pen*, a streamlined pen with pointed ends, which was to become the most prominent fountain pen form for decades to come.

In 1931 came the *Feather Touch*, that glided gently over the paper without pressure. The platinum-tipped nib had a special slit that made it possible to write in wide or narrow lines by turning the pen. To meet the growing demand for reasonably priced pens, Sheaffer offered the *Wasp* from 1934 on, at prices that began at $1.95. This series had a transparent ink reservoir. For the sake of the firm's image, the name of Sheaffer was not used on them. Only the reference to the place of manufacture, Fort Madison, let one draw conclusions as to the well-known manufacturer. The tubeless filling system of the *Wasp* was also used for all other Sheaffer fountain pens as of 1935. The transparent barrel let the user check the ink level conveniently. In 1936 the Sheaffer *Lifetime* appeared in a new form. Instead of being marbled in color or in one solid color, these pens now bore elegant shimmering stripes in green, golden brown, gray-black, or red. In 1939 Sheaffer introduced the *Crest,* one of the first fountain pens to have a metal cap. For those who were fighting in World War II, in 1940 Sheaffer produced pens whose clips extended over the end of the cap. They could be carried more easily in uniform pockets that way.

Like most of his competitors, Sheaffer also had to cut production during World War II and manufacture equipment and armaments for military use. Craig Sheaffer, who had succeeded his seventy year-old father at the head of the business in 1938, was nevertheless able to continue pen development to a limited extent. As a result of these efforts, the *Triumph* was created in 1942. The cylindrical, conical-ended pen of 14-karat gold had an aura of great value. Like previous pens, the *Triumph* was also available as the *Tuckaway*, without a clip, for ladies, or with a very short and barely flexible clip.

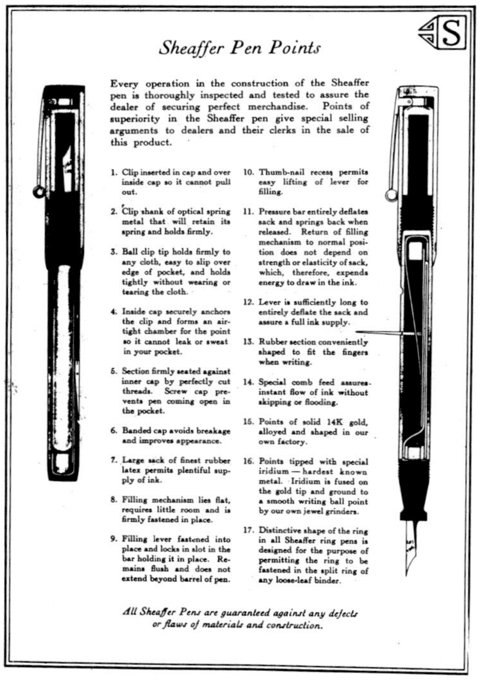

Sheaffer Pen Points

Every operation in the construction of the Sheaffer pen is thoroughly inspected and tested to assure the dealer of securing perfect merchandise. Points of superiority in the Sheaffer pen give special selling arguments to dealers and their clerks in the sale of this product.

1. Clip inserted in cap and over inside cap so it cannot pull out.
2. Clip shank of optical spring metal that will retain its spring and holds firmly.
3. Ball clip tip holds firmly to any cloth, easy to slip over edge of pocket, and holds tightly without wearing or tearing the cloth.
4. Inside cap securely anchors the clip and forms an air-tight chamber for the point so it cannot leak or sweat in your pocket.
5. Section firmly seated against inner cap by perfectly cut threads. Screw cap prevents pen coming open in the pocket.
6. Banded cap avoids breakage and improves appearance.
7. Large sack of finest rubber latex permits plentiful supply of ink.
8. Filling mechanism lies flat, requires little room and is firmly fastened in place.
9. Filling lever fastened into place and locks in slot in the bar holding it in place. Remains flush and does not extend beyond barrel of pen.
10. Thumb-nail recess permits easy lifting of lever for filling.
11. Pressure bar entirely deflates sack and springs back when released. Return of filling mechanism to normal position does not depend on strength or elasticity of sack, which, therefore, expends energy to draw in the ink.
12. Lever is sufficiently long to entirely deflate the sack and assure a full ink supply.
13. Rubber section conveniently shaped to fit the fingers when writing.
14. Special comb feed assures instant flow of ink without skipping or flooding.
15. Points of solid 14K gold, alloyed and shaped in our own factory.
16. Points tipped with special iridium — hardest known metal. Iridium is fused on the gold tip and ground to a smooth writing ball point by our own jewel grinders.
17. Distinctive shape of the ring in all Sheaffer ring pens is designed for the purpose of permitting the ring to be fastened in the split ring of any loose-leaf binder.

All Sheaffer Pens are guaranteed against any defects or flaws of materials and construction.

The advantages of the Sheaffer design, as described in the 1925 catalog.

Sheaffer was one of the first manufacturers of writing instruments to include the new ballpoint pen in their offerings of 1946. The *RAI Stratowriter* was sold in a gold-plated metal case, and, pointing the way to the future, already had push-button operation. The stylishly slim TM (Thin Model) line was provided with a new filling system in 1950: *Touch Down*. Its development went back to the beginnings of fountain pen history. After unscrewing a handle at the end of the barrel, a piston rod was pulled out and pushed back into the pen to fill it. To improve this simple and easy filling method, Sheaffer created the *Snorkel Pen* in 1952. Now the user moved a rod like a submarine periscope out from under the nib and dipped it into the ink to fill the pen. Since the nib no longer touched the ink in the bottle, there was no need to wipe ink off it.

Despite this advantage, filling the pen from an open ink bottle still deserved to be improved on. Sheaffer recalled an idea that dated back to the turn of the century. In 1955, low-priced plastic ink cartridges were mass-produced and sold under the Sheaffer name.

In 1959 Sheaffer produced unusually big, heavy and compact fountain pens for his largest group of customers, the male purchasers. The *PFM* (Pen for Men) had proven construction features like the *Snorkel* and *Touch Down*, and also offered a spring clip to hold fast to the shirt pocket, a metal cap, and a new type of nib inlaid in its mounting in intarsia style. The striking, harmoniously designed pen soon came to symbolize Sheaffer pens. Developed further in the modern *Targa* series, it is still produced today.

The tradition-rich firm of Sheaffer turned back to its past successes in the 1970s: The *No-Nonsense* series of writing instruments, whose styling was based on Sheaffer pens produced from 1915 to 1918, are available as fountain pen, ballpoint pen, fineliner and calligraphy pen and rank among the writing sets that have won friends all over the world. In 1970 the high-priced *Nostalgia* series followed, with silver and gold filigree recalling the days of Youth Style, and in 1986 the *Connoisseur* joined them. Today again, fountain pens accurately reproducing the originals from the "roaring twenties" are successfully being offered to a very interested public.

The cylinder-shaped Lifetime pen of 1944.

A romantic advertising theme of 1920.

"Always writes"—decorative fountain pens and mechanical pencils of 1920.

Torpedo-shaped "Balance" writing instruments of 1929.

The Fountain Pen—

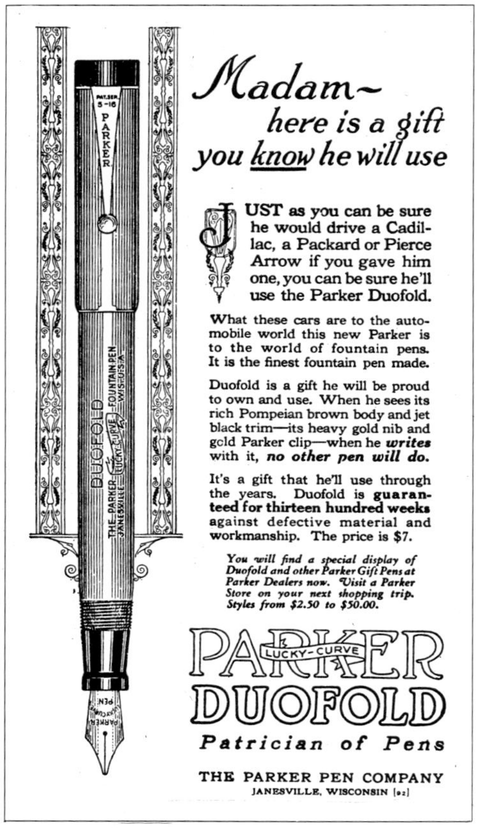

Comparison with an expensive car—the oversize Parker Duofold of 1921.

Waterman was the first to produce hard rubber in color.

The number of fountain pens and mechanical pencils available on the international market in the 1920s was incalculable. Not only in styles, choice of ornamentation, or variety of filling systems, but also in the number of available nib widths, there were simply too many models on the market. Parker alone offered an assortment of four hundred models in 1921, all different but hardly to be told apart by the layman. The average price for a Parker fountain pen was about $3.50, a lot of money for the typical American in the years after World War I. George S. Parker naturally disagreed when one of his employees suggested making a new fountain pen to hitherto unknown dimensions. The new model should be bigger, thicker, and twice as expensive as was customary before. "Look out the window, Mr. Parker, and observe the cars on the street. Every year brings new models, bigger, heavier, and more expensive. And the people buy them. Why shouldn't they want to buy a pen for seven dollars if they can be proud of a big car?"

The argument had its effect. The first test models were almost ripped out of Parker's hands. In 1922 the time had come: The Parker *Duofold*, an extra-large fountain pen with an extra-large nib and extra-large ink supply, attracted the public's attention. Equipped with an elegant "Pompei brown" lacquered-china type of finish, the *Duofold* fascinated a public who could afford it. In the first year Parker's turnover increased by 77%, and in three years it quadrupled. Parker concentrated completely on the new model and reduced the assortment from some four hundred models to barely thirty.

Aside from the metal parts such as the nib, clip, and mechanism, hard rubber, formerly called ebonite or vulcanite, was used to make fountain pens. Ebonite was the only material of its time that could be shaped and worked easily, was scarcely permeable by water or air, and did not change its shape, even under great temperature and humidity variations. In the first twenty years of this century, the fountain pen's world was correspondingly monotone, completely black almost without exception, and at certain price levels decorated with contrasting ornaments and filigrees of silver or gold alloys in the fashion of the Victorian era, Youth Style, or Art Deco.

The L.E. Waterman Company tried to get away from the masses of uniform pens. In 1923 that firm succeeded in producing a colored caoutchouc mixture that let the cap and barrel shine in red, yellow, and green "flaming" finishes. The public was charmed. Colored fountain pens became the fashion.

Their competitor Sheaffer had been working since 1910 on a policy of producing fountain pens in fashionable colors. More ambitious than his competitor Waterman, he also tried to eliminate one disadvantage of hard rubber pens, the ease with which the sensitive natural material broke. It had to be a new kind of material. Sheaffer put a lot of time and money into his project, but it failed. The synthetic material so laboriously developed expanded too much under heat.

Help came from another side. The chemical firm of Dupont developed the cellulose material "Pyroxylene" to production readiness in 1924. It could be made in tubular form, could be worked and shaped neatly and produced in hundreds of colors and patterns—striped, marbled, speckled, shimmering like mother-of-pearl. And it also had the great advantage of being completely unbreakable and undentable.

Sheaffer was the first manufacturer of writing instruments to use it, and had his customers' enthusiasm working for him when he added a lifetime guarantee to his pens made of "Radite," an imaginative name for Pyroxylene. Sheaffer's *Life-Time Pens* bore a white dot on the cap, over the clip, since 1924 as a mark of quality.

The Golden Years

The golden years of the fountain pen were also the golden years of the big, famous names. In America, and in England and France as well, the list of prestige names was led by Waterman, followed by Parker and Wahl-Eversharp. Wahl, a calculating-machine manufacturer founded in 1905, won public acclaim with a new type of mechanical pencil (the always sharp pencil: Eversharp) invented by a Japanese industrialist, and went into fountain pen production in 1918.

In England, Thomas de la Rue & Co. Ltd. reached the top with the "Pelican"—the first models were made in London in 1897—and "Onoto" brands. (Pelican is not to be confused with the German brand of Pelikan). The great brands of Swan (Mabie, Todd & Co.) and Onoto (Thomas de la Rue), plus Conway Stewart, were the most successful English trio.

Germany, the land of poets and thinkers, did not take up the American inclination to use the convenient fountain pen instead of the simple steel pen until relatively late. The products offered by the tradition-rich steel pen firm of Soennecken began to include insertion fountain pens around the turn of the century, and also safety fountain pens as of 1906, but they played a more or less unimportant, "outsider" role. In the 1920s there came lever fillers, in the 1930s pressure fillers like the famous *President* and *Rheingold*, as well as piston fillers.

The pencil manufacturer A. W. Faber offered a popular assortment around 1900. His steel pens, fountain pens, and ink, produced in his branch factories in Noisy-le-Sec outside Paris and in New York, were intended mainly for export markets. As one of the leading European brands, A. W. Faber sold not only insertion fillers but also a safety fountain pen. After World War I, the firm really got into the fountain pen picture again when it acquired the "Osmia" pen factory.

The "Heidelberg Penholder Factory of Koch, Weber & Company" ("Kaweco") was an important producer of writing instruments and made a significant contribution to the history of their development. Over a hundred different fountain pens, steel pens and other types made up their line even before World War I. Kaweco praised their safety pen with "a justifiable right to call it the most perfect fountain pen of the present" (1911). and thus made it the typical European fountain pen of the first two decades of our century.

The Hamburg Simplo-Filler Pen Company operated not only in Germany but also in England and France, offering a considerable range of high-quality "Rouge et Noir" safety pens, "Simplo" insertion pens, and indelible pencils. Its trade mark was a five-pointed white star with the highest mountain in Europe, Mont Blanc, as a symbol of the highest quality. Of the three typical Montblanc price ranges, the "low-price level" (with 300 numbers, Reflex), "medium level" (200 numbers), and "masterpiece class" (100 numbers), which held the highest rank from 1924 on. These large volume, impressive pens in black, blue, green, and pearl soon ranked among the top international fountain pens. Montblanc used almost all the proven filling systems in its ample line during the 1930s: safety pens, lever, push-button, pressure, and control fillers.

As one of the last but no less successful competitors, Günther Wagner entered the field in 1929 with the "Pelikan" piston-filler pen. Günther Wagner enjoyed an excellent reputation far beyond the borders of Germany for the quality of his paintbrushes, office supplies, and world-famous 4001 inks. This image of quality could also be extended to the later public favorite, the dark green Füller 100. Astonishingly, the Pelikan pen based its success on only one type: the Model 100 in a variety of colors and precious-metal versions (the *Rappen* and *Ibis* special models played only a secondary role).

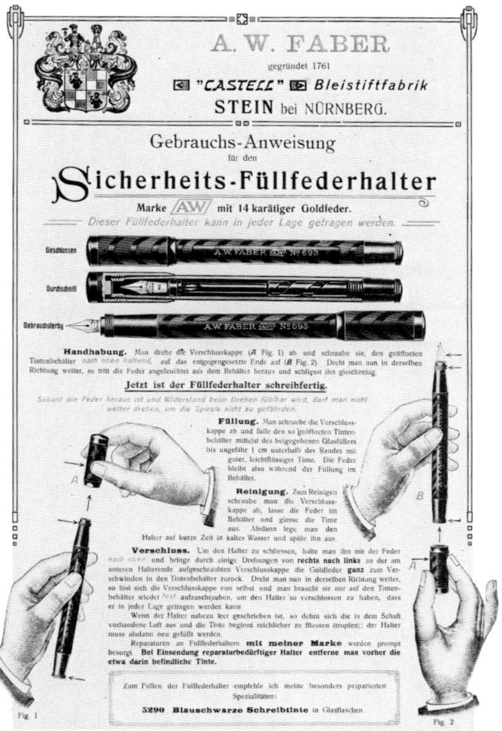

A. W. FABER
gegründet 1761
"CASTELL" Bleistiftfabrik
STEIN bei NÜRNBERG.

Gebrauchs-Anweisung
für den
Sicherheits-Füllfederhalter
Marke AW mit 14 karätiger Goldfeder.

Dieser Füllfederhalter kann in jeder Lage getragen werden.

Geschlossen
Durchschnitt
Gebrauchsfertig

A.W. FABER AW № 693

Handhabung. Man drehe die Verschlusskappe (*A* Fig. 1) ab und schraube sie, den geöffneten Tintenbehälter *nach oben haltend*, auf das entgegengesetzte Ende auf (*B* Fig. 2). Dreht man nun in derselben Richtung weiter, so tritt die Feder angefeuchtet aus dem Behälter heraus und schliesst ihn gleichzeitig.

Jetzt ist der Füllfederhalter schreibfertig.

Sobald die Feder heraus ist und Widerstand beim Drehen fühlbar wird, darf man nicht weiter drehen, um die Spirale nicht zu gefährden.

Füllung. Man schraube die Verschlusskappe ab und fülle den so geöffneten Tintenbehälter mittelst des beigegebenen Glasfüllers bis ungefähr 1 cm unterhalb des Randes mit guter, leichtflüssiger Tinte. Die Feder bleibt also während der Füllung im Behälter.

Reinigung. Zum Reinigen schraube man die Verschlusskappe ab, lasse die Feder im Behälter und giesse die Tinte aus. Alsdann lege man den Halter auf kurze Zeit in kaltes Wasser und spüle ihn aus.

Verschluss. Um den Halter zu schliessen, halte man ihn mit der Feder *nach oben* und bringe durch einige Drehungen von **rechts nach links** an der am unteren Halterende aufgeschraubten Verschlusskappe die Goldfeder **ganz** zum Verschwinden in den Tintenbehälter zurück. Dreht man nun in derselben Richtung weiter, so löst sich die Verschlusskappe von selbst und man braucht sie nur auf den Tintenbehälter wieder *fest* aufzuschrauben, um den Halter so verschlossen zu haben, dass er in jeder Lage getragen werden kann.

Wenn der Halter nahezu leer geschrieben ist, so dehnt sich die in dem Schaft vorhandene Luft aus und die Tinte beginnt reichlicher zu fliessen (tropfen); der Halter muss alsdann neu gefüllt werden.

Reparaturen an Füllfederhaltern **mit meiner Marke** werden prompt besorgt. **Bei Einsendung reparaturbedürftiger Halter entferne man vorher die etwa darin befindliche Tinte.**

Zum Füllen der Füllfederhalter empfehle ich meine besonders präparierten Spezialitäten:

5290 Blauschwarze Schreibtinte in Glasflaschen.

Fig. 1 Fig. 2

A.W. Faber instructions for the safety filler, 1908.

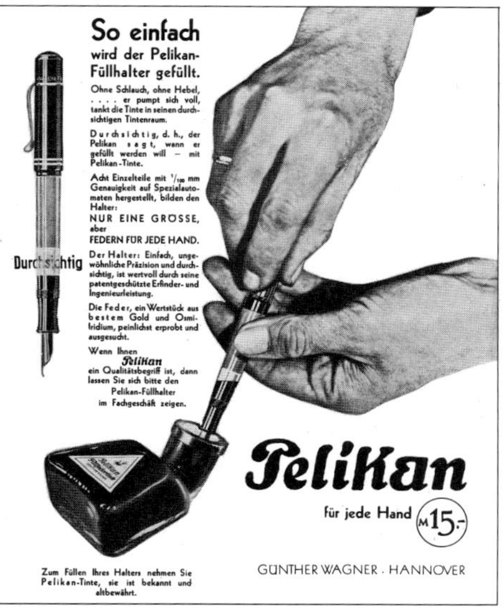

A piston filler in one size, but pens for every hand, were made by Pelikan as of 1929.

In Italy the firms of OMAS and Aurora were the most famous names in fountain pens. OMAS (Officine Meccaniche Armando Simoni) was founded in 1919 as a small repair shop for English and American fountain pens. Five years later the former watchmaker Armando Simoni produced his own fountain pen, a special model for doctors with a built-in fever thermometer. In the 1930s the twelve-sided OMAS *Lucens Extra* pens, with arrow clip, transparent ink reservoir, and an arrow on the nib, ranked among the great successes of the Italian firm—as they still do today.

In France there were, along with Jif-Waterman, smaller manufacturers such as Conté, Mallat, Bayard, and Paillard, trying to compete against American, English, and German imports.

In Japan in the 1920s, imported models remained the favorites for a long time, with Onoto and Swan ranking among the best-known. Among native manufacturers, the Pilot Pen firm, founded in 1917, and the Platinum, established in 1919, gained significance in the golden 1920s.

The German Melbi fountain pen with lever mechanism and glass point, circa 1930.

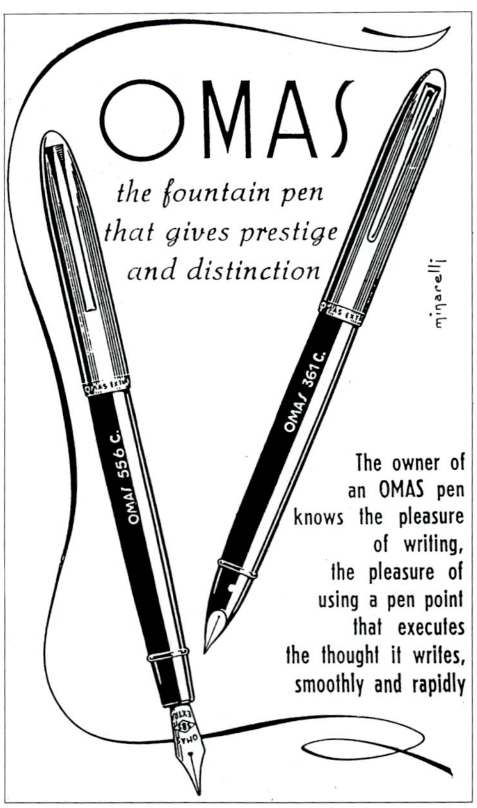

The great Italian success. A 1955 advertisement.

Montblanc stressed the white vest in this placard, circa 1920.

The world's leading fountain pen market, North America, grew and grew. Whereas in 1905 barely one and a half million pens were sold, the sales figures rose to twenty million in 1929. The variety of colors and styles was practically endless. The classic cylindrical form was regarded as the only shape for a fountain pen. For a long time it had lost its precious-metal decorations from the Victorian and Youth Style eras. It presented itself as simple, elegant, and understated, the form that we regard as classic for the fountain pen today.

Along with the typewriter, which had come to be taken for granted in offices, the fountain pen was the preferred writing instrument for the one who had something important to write: the boss. To get the pen into his hand faster, and to meet the need for something with more class, around 1915 the great manufacturers of writing instruments began to offer penholder sets to place on the desk. They replaced the heavy metal, wood, or marble sets that had been customary until then, and that had included one or two inkwells with closing tops and a groove in which to lay the pen. Along with Parker and Sheaffer, Wahl-Eversharp ranked among the manufacturers who offered a large line.

With the new decade came a change of taste, beginning in America. The angular classic writing instruments were reborn in streamlined form. While the market leader, Waterman, still delayed and let almost ten years pass before just barely following the trend, Parker made use of the new fashion from the start. As of 1929, Parker's caps and barrels took on new forms.

The most obvious design change of the 1930s was made by Wahl-Eversharp in 1931. The *Doric* pen, in pure Art Deco style, was an important change in the design of fountain pens. Based on the classic example of Greek columns with their elegant, flowing lines, the *Doric* was given a twelve-sided barrel and cap surface. Along with colors that suggested the Orient, this pen reached a high point of design in the aesthetics of writing instruments in its day.

Sheaffer's writing instruments, with their "Balance look," were the ultimate in the more and more popular streamlined style. Shaped like a cigar, almost without angles or notches, they fit smoothly into the writing hand. With this pen, Sheaffer found its characteristic form in 1929, which was retained until the 1970s with only minor variations.

In 1933 a new star appeared in the heavens of the pen world: the revolutionary *Vacumatic* filling system was born. Developed by a scientist at the University of Wisconsin under contract with Parker, it offered a maximum in easy use and a larger ink reservoir.

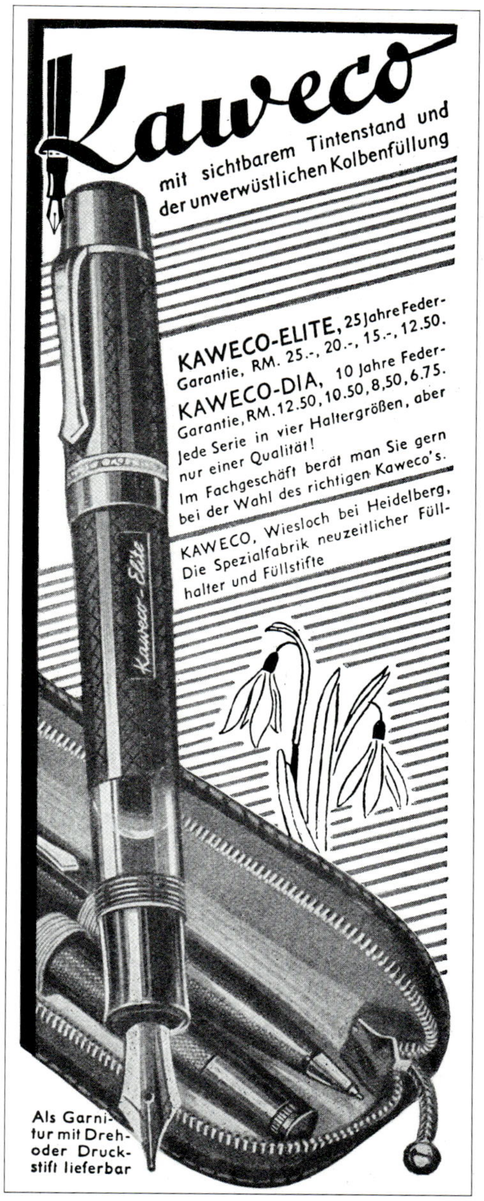

Kaweco-Elite: a modern 1938 fountain pen with 25-year guarantee.

The "Senator" fountain pen with push-button mechanism, circa 1940.

Mounted on horseback came the "Stylos" of Bayard in 1922.

For a whole decade, the *Vacumatic* remained one of the most significant technical developments in the history of the fountain pen.

Aside from the refinement of the piston mechanism, which Pelikan, Montblanc and other European firms had achieved, the fountain pen manufacturers limited themselves, as in the early years of the fountain pen, to design and fittings. At the beginning of World War II, the further development of the fountain pen in Europe came to a stop, and to some extent died completely. Instead of gold nibs, only steel ones were allowed.

Again it was Parker who established a new milestone. In 1941 there appeared a writing instrument whose design had practically nothing in common with previous fountain pens, a truly different, futuristic design. It bore the characteristic name of "Parker 51," probably to indicate that it had been planned for ten years to come. Only the *Vacumatic* filling system and the Parker arrow clip resembled those of its predecessors. The most important and interesting changes were embodied by the cap, which for the first time was made of metal and no longer screwed on, but simply slid on and was held by a thin metal ring in the barrel, as well as the nib that was almost completely set into the streamlined front section. Another important advantage: blotting the ink, according to the advertisements, was a thing of the past. Parker offered an ink that penetrated the paper very quickly and did not smear, and could advertise proudly: "Writes dry with wet ink!"

At the beginning of the 1940s, Sheaffer contested the top position in the North American market with Parker. To slow Parker's sales success with the "51," Sheaffer brought out the *Triumph Lifetime Pen* in 1942. This writing instrument ranked among the best ever in terms of material and quality. Instead of having a normal, gently arched nib, the *Triumph* offered a huge cylindrical gold tip. Filling was done by a vacuum system similar to Parker's. To be sure, the *Triumph* was also available as a lever or piston filler. A special inkwell made filling easier. By tipping when closed, the ink was moved into a scoop in the neck of the bottle. This trick, like the famous lie-down bottles of Pelikan and Montblanc, allowed filling without getting ink on the front of the pen and the fingers.

The third American competitor, Wahl-Eversharp, attracted the buyer's attention in 1941 with a new and different cap design. Aimed entirely at the cities, which presumably symbolized most strongly the impulse toward progress that emanated from mighty America, the name of Eversharp's new series was taken from the profile of New York City: *Skyline*. The big cap on the delicate barrel, which almost came to a point at the end, had an imposing effect. The slim clip in band form, unlike those of previous models, curved over the rounded head of the cap. Along with a fantastic pen, Eversharp had one more special feature to show: The "magic" ink passage prevented the ink from running out and spattering, even at high altitudes. It was, therefore, one of the first fountain pens to be "flight suitable," a quality that was appreciated by the Air Force pilots and that helped it to charm the buying public during the wartime years. The *Skyline* series developed into a remarkable success. Along with Eversharp's 1944 marketing and intriguing radio advertising, it quickly made Eversharp the most successful marketer of writing instruments in the USA in 1945.

A change in the sense of shape: Wahl-Eversharp's "Doric" line, introduced in 1931.

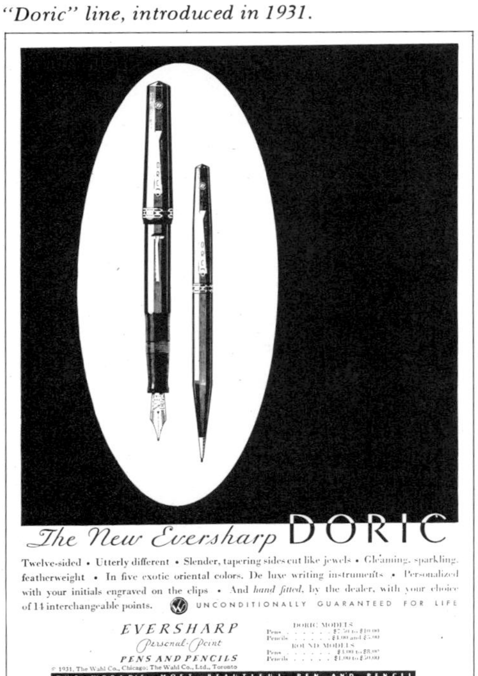

The fourth of the great American pioneers in fountain pen history, Waterman, threatened to fall farther and farther behind in the competition for the buyer's favor. With the introduction of the *100-Year* fountain pen in 1939 the firm was able to stop its fall for a short time. This unusual series is still regarded by connoisseurs and collectors today as one of the most fascinating writing instruments of all time. The streamlined cap and barrel consisted of semi-transparent colored plastic with its surface broken neatly by encircling grooves. It was crowned by inset ends of the cap and barrel, made of glowing transparent plastic in a contrasting color. But Waterman was not able to make use of its last great public success in America.

Presentation of Writing Instruments

"What the customer does not see, he does not buy!" Rule Number One for every good merchant did not matter much in the early years of the fountain pen. The stationery dealer carried a line of fountain pens, still regarded by some as exotic, as a sideline. Paper, artists' paints, and office supplies were the main merchandise. The fountain pen was pulled out of a drawer only when it was specifically requested. So as not to fall behind one's competitors in a tough market, something had to be done. Smart advertising meant primarily smart presentation.

For seven shillings sixpence the "all British" manufacturer Conway Stewart "made writing a pleasure" in the 1920s.

With an order for a suitably large number of fountain pens, Kaweco delivered this "very elegant sales case for 24 fountain pens" in 1911.

Ink-Kuli display from the 1950s.

"This sales case for 48 Soennecken fountain pens is arranged very usefully and makes sales much easier. Made of fine crystal glass with a base of finely polished mahogany, and lined with fine plush. An ornament for the finest store counter." 1927.

This porcelain display of 1938 bore seven Pelikans

Wahl-Eversharp

This great American manufacturer of writing instruments is the offspring of an unlikely pair of parents. In 1905 John C. Wahl founded the "Wahl Adding Machine Company" and produced calculators and mechanical pencils. In 1912 the Japanese, Tojuki Hayakawa, invented a mechanical pencil and founded an "Eversharp Pencil" company. Out of them, in the course of decades, there developed the huge "Sharp" electronic corporation. Hayakawa exported his "Eversharp" pencils to the USA with great success. In 1915 the Wahl Adding Machine firm bought Hayakawa's American branch and began its own production of the Japanese mechanical pencil. The Americans seem to have been as interested then as they are now in good products made in Japan. In any case, the demand was so great that Wahl-Eversharp decided to sell their adding machine manufacturing facilities to Remington in 1915, buy the Boston Fountain Pen Company, and specialize in fountain pens and mechanical pencils. The firm's name was thus shortened to "Wahl Company." Their first hard-rubber fountain pens, made in 1918, were called *Tempoint,* and in only two years 1.5 million of them per year were being sold. Within a short time Wahl had pushed his competitors, Parker and Waterman, down to second and third places.

When the old hard-rubber pens were replaced by those of the modern Dupont pyroxylene plastic, Wahl again had the opportunity to rise to the leading position among America's producers of writing instruments. The firm's success could even be reinforced in 1931 with the introduction of the *Doric,* a twelve-sided series of pens designed to resemble the doric columns of ancient Greece. A golden seal, like Sheaffer's white dot, promised a lifetime guarantee.

In 1932, Wahl equipped the popular Doric with an additional feature. The pen could be adjusted nine ways in terms of width and flexibility.

In 1941 came the best-selling pen series in the history of the Wahl-Eversharp firm: the *Skyline.* This unusually designed fountain pen had an asymmetrical shape. The optical center of attention was the cap with its "military" clip extending over the end of the cap, a necessity in the 1940s for soldiers, sailors, and airmen. The end of the barrel tapered almost to a point. The new *Skyline* doubled Eversharp's (the name was changed in 1940) income within a year. A thorough advertising campaign made the new series even more familiar and popular. Practically every child knew the name of Eversharp from a radio quiz program. For correctly answering the "sixty-four-dollar question" the candidate received an Eversharp pen-and-pencil set with the number "64" engraved on the massive gold cap. The same radio commercials were used for the *Fifth Avenue,* a sister series of the *Skyline.* It was cylindrical in form and conservative in style, but could not match the success of the *Skyline.*

In 1945, the firm's future could not have looked brighter, even if Eversharp had not joined Eberhard Faber in acquiring the exclusive rights of Biro, the inventor of the ballpoint pen, from the Eterpen firm of Argentina. But then misfortune appeared. Shortly before the "Capillary Action" ballpoint pen, further developed by Eversharp at a high cost, could be put on the market, and unknown man named Reynolds entered the competition as the first to offer a ballpoint pen in the USA. Determined to give their opponent a fight, Eversharp produced and marketed its not completely refined ballpoint pen without regard for the quality and the excessive cost of machinery and equipment. The result was disastrous. In 1948 the firm lost millions. The Eversharp firm, once so successful, could only avoid the threat of bankruptcy by using the growing profits of its Schick razor branch to offset the writing instrument branch's losses. Sales sank, personnel had to be discharged. A last attempt with the *Ventura* series in 1950 could not prevent the firm's collapse. In 1957 its competitor, the

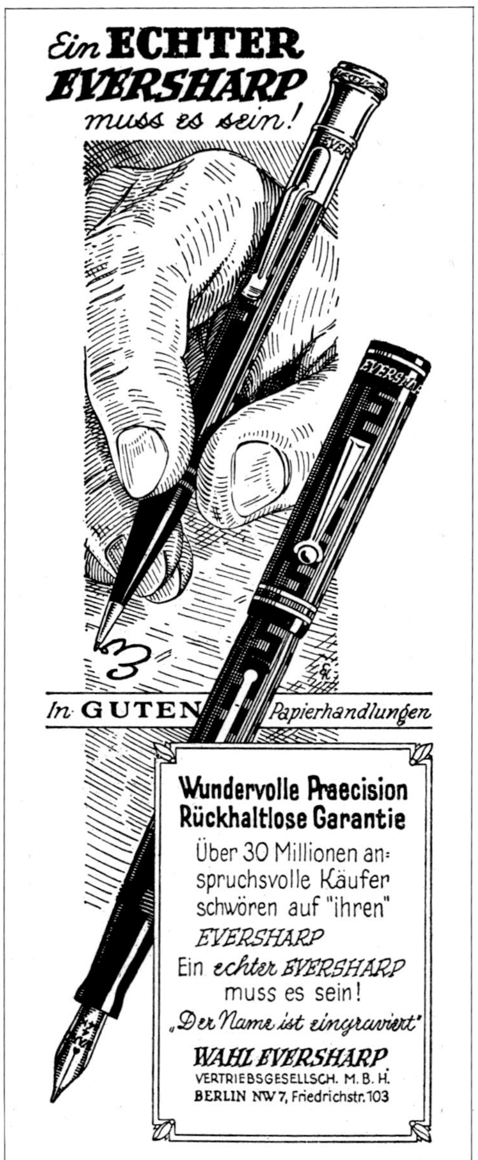

The "Eversharp" mechanical pencil, invented by the founder of the huge Japanese Sharp firm. The early Wahl "Tempoint" lever fillers already had the typical Wahl clip with a movable ball. A 1928 advertisement.

Parker Pen Co., bought the Eversharp writing instrument division with the intention of being better able to cover the lower price ranges. After initial success with reasonably priced ballpoint pens, Parker gave up this double strategy. Eversharp continued to cause losses. One of the pioneers of the industry had to give up once and for all in 1959.

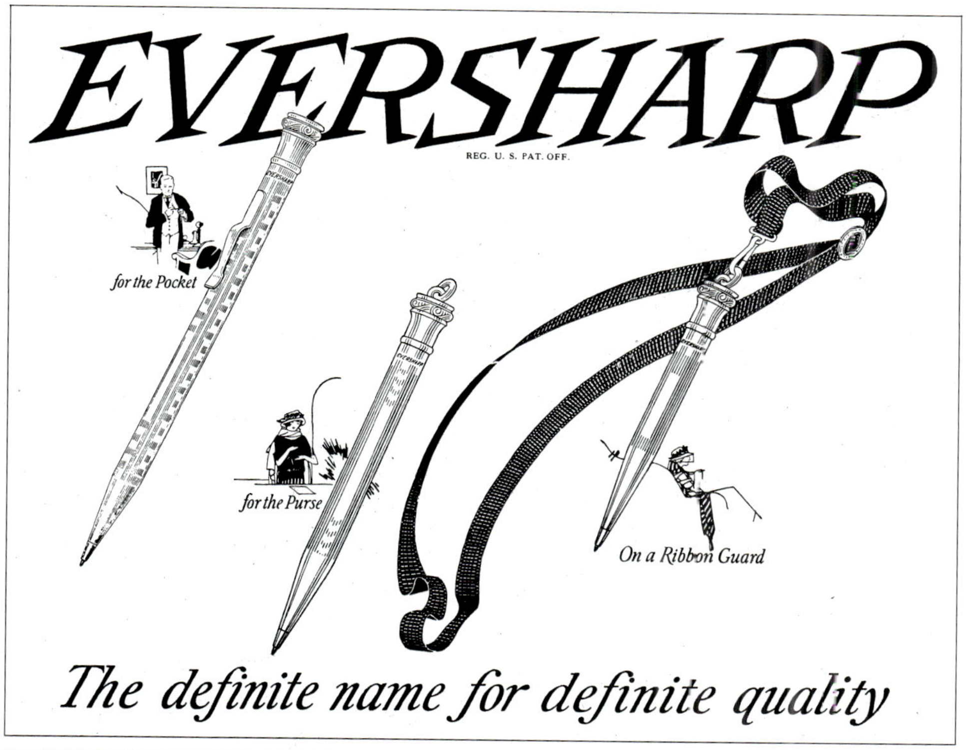

With always sharpened and thus "eversharp" mechanical pencils for every use, the firm's progress began.

At the height of success: The 1942 "Eversharp Skyline." The popular radio quiz program with the $64 question quickly made Eversharp popular.

Kaweco

Kaweco is one of the oldest German fountain pen manufacturers. Founded in Heidelberg in 1897 by two merchants, Rudolf Weber and Otto Koch, it developed from a small beginning into the "Heidelberg Penholder Factory of Koch, Weber & Co." ("Kaweco" for short), one of the most important firms in its field in the first twenty years of our century. By 1911 there were already branch offices and repair shops in Paris, Zürich and Vienna. Kaweco had representatives in almost all the capitals of Europe, including St. Petersburg, and also in Cairo, Istanbul and Oran.

At this time Kaweco could claim to offer the most inclusive and best-quality line of fountain pens. Along with the time-honored immersion filler with conical or cylindrical cap, brand name *Perkeo*, there was the *Swallow*, an American product purchased from A. Morton of New York, with "upper and lower ink passages." The Kaweco catalog also included pendant fountain pens, pens for notebooks and briefcases, combination pens with two pens and two caps for red and blue ink, as well as pens with gold ink passages. Specific attention was called to "older fountain pen systems" (the text ran: ". . . in their place, we recommend without exception our Kaweco safety fountain pens as up-to-date and more purposeful"). There were piston fillers, tube fillers with side apertures to press the tube together, and pressure fillers.

All attention was directed to the new Kaweco safety fountain pen. The advertisement sounded self-conscious: "Wherever it has become known, all other types have been rejected and it has been preferred. Its form has been imitated by almost all manufacturers inside and outside the country. It has truly conquered the world in flight." This lets one conclude that Kaweco belonged among the pioneers of the popular safety pen along with Caw, Soennecken, Waterman, and Faber-Castell.

But its pioneering seemed not to have been limited to the safety pen. As a fountain pen factory, Koch, Weber & Co. concentrated not only on the fountain pen, but also on other writing instruments with liquid ink. The *Stylo* ink writer, which had a metal tube with an adjustable needle instead of a nib, was also produced, as was, surprisingly, the *Signier* fountain pen. This may well have been the oldest known European felt-tip pen. The discovery of the felt-tip pen is generally ascribed to the Japanese. It is scarcely known that Kaweco was selling a fillable marker for placards, cases, packages, and sacks. It was equipped with a felt tip as a writing nib, which could be made round or flat in varying widths. Famous personalities like Count Zeppelin stated that "the Kaweco fountain pen works to my satisfaction, for long times too."

New competitors and decreasing orders from abroad led to increasing financial difficulties, which had to be warded off in 1927 with a forced compromise with the firm's creditors. Two years later the formerly successful business, that had employed 160 employees, had to declare bankruptcy. Some of the machines and supplies, as well as the brand name, were taken over by the "Baden Fountain Pen Factory of Knust, Worringen & Grube," a firm founded in 1925. This small company, employing about eighty people, had already gained a good name with models like the *Aurumia-Original*, *Aurumia-Spezial*, and *Kollek*.

Under the direction of Friedrich Grube, a new upswing began. Colorful marble-finished piston and tube fillers appeared under the Kaweco name, and the line was expanded by pressure pencils. A sales hit developed from an idea that Koch, Weber & Co. had developed around 1910 as a pen for "ladies, officers, and sportspersons": the short, handy *Kaweco-Sport* fountain pen and mechanical pencil. These sets in their leather cases were suited, on account of their pocket format, to "sport, service, and social dress."

Kaweco ranked among the leading German fountain pen brands at the beginning of the 20th century.

After World War II, Kaweco concentrated on two types of writing instruments: the *Dia* piston fountain pen with transparent ink reservoir (Models 85a and 90a) and the *Elite* (183a, 187a and 190a, the last for ladies). But the tried and true *Kaweco-Sport* line also experienced a renaissance. The hexagonal prewar model was rounded and equipped with a modern type of gold nib in the early 1960s.

When Friedrich Grube died, his sons took over the direction of the Kaweco brand. The *Kaweco, Monopol, Kadett,* and *Kolleg* fountain pens, ballpoint, and mechanical pencil models, though, could no longer get by on their pre-war success. Wilhelm Grube, the sole proprietor since 1964, had to struggle more and more against hard times. In 1970, the family business, whose brand name had originated in the previous century, declared bankruptcy. A successor firm, directed by Liselotte Grube since 1970 and employing fifteen workers, has operated in modest circumstances near Heidelberg.

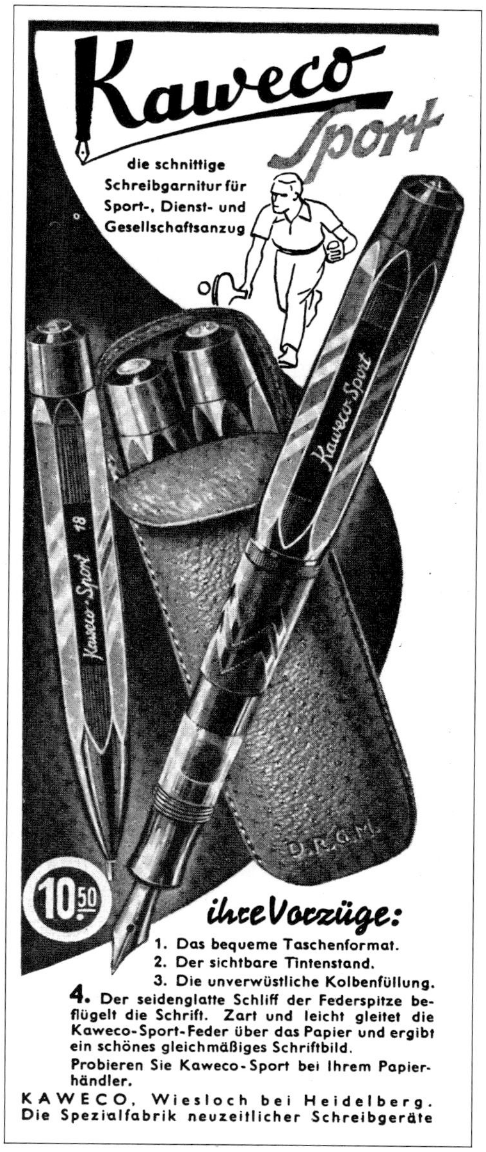

The sales hit of 1938: the short, compact "Sport" fountain pen.

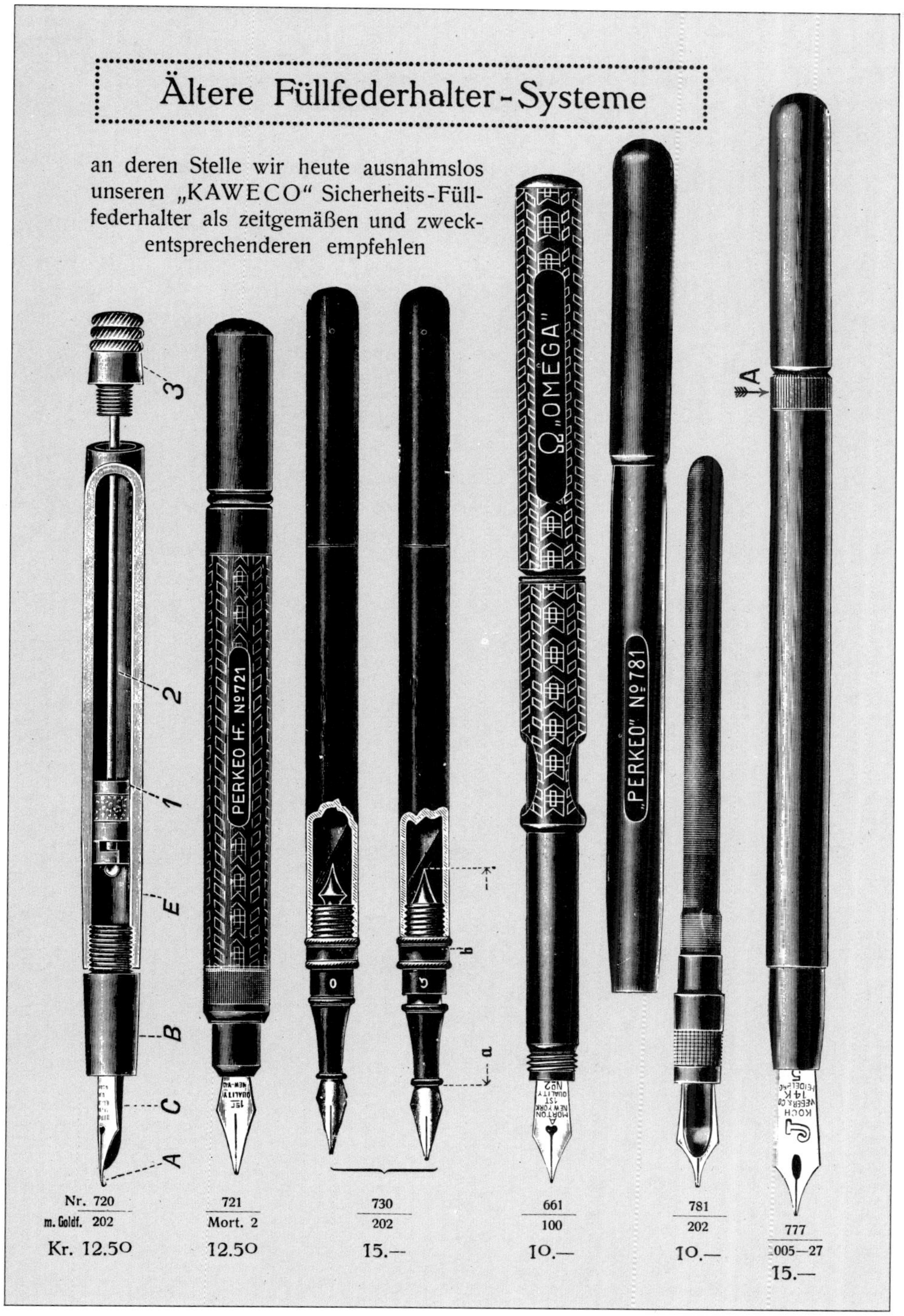

Older Fountain Pen Systems
in place of which we recommend today, without exception, our "KAWECO" safety fountain pens as timely and purposeful.

Unusual for modern-day advertising: Older models were specifically not recommended, for the sake of a new product.

Montblanc Simplo

When Carl Schalk won a second game of cards on that momentous evening in 1910, the name was born: "Call the new fountain pen Montblanc. It is also black on the bottom, white on top, and the greatest among equals . . ." And so it was done. For this close relative of the Voss family of Hamburg could not have expressed the standards of quality of this rising new pen company better. The round shape of the pen's cap, which was originally white, was turned into a star to represent the peak of the Montblanc Massif. The height of this mountain massif, 4810 meters, was engraved on the nib of this costly fountain pen by its makers in Hamburg.

The history of the Montblanc firm had begun with an agreement made in 1906. A Hamburg stationery dealer agreed to satisfy the demand for the new fillable pens with his own products. A Berlin engineer took charge of manufacturing, and a Hamburg banker handled the finances. This three-man team was successful beyond their expectations. In 1908 they founded the Simplo-Filler-Pen-Company of Max Koch for the purpose of attracting a large number of customers.

Circa 1910: From the beginning, Montblanc was an international thinking business.

A humorous placard design from the 1930s.

The originally silent partner Claus-Johannes Voss functioned as the new financier. In the next two years Christian Lausen and Wilhelm Dziambor also became partners; their families were to direct the Montblanc firm for nearly seventy-five years and be regarded as the actual founders or founding families.

The headquarters of the firm remained in Hamburg, the "gateway to the world," the location being advantageous in terms of facilitating export activity. Success continued. At first the firm limited itself to the production of barrels and caps for insertion and safety pens of the *Rouge et Noir* brand. Anything French was regarded then as especially attractive. Only later was the typical red cap replaced by a white one which, after Wilhelm Dziambor's famous card game in 1913, was changed into a star.

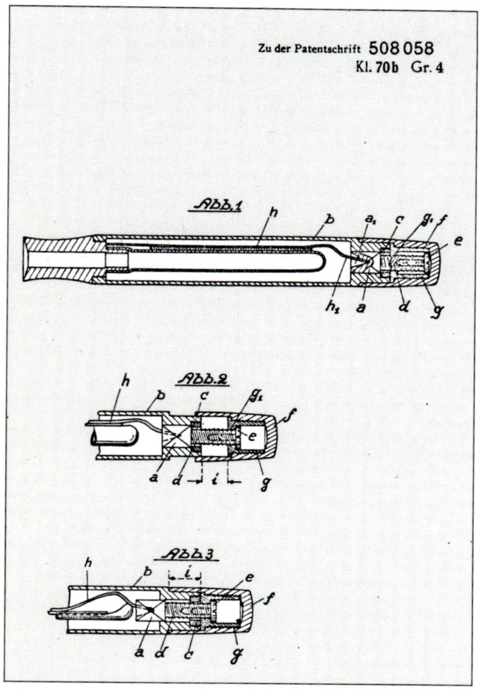

For the push filler, Montblanc had an original system patented: To fill it, one unscrewed the filling grip on the barrel and pressed it. No end cap could get lost, unlike those of pressure fillers.

A 1908 airplane placard.

The gold pens of this era were still made in America. Having meanwhile become famous for their quality, they were first produced in Hamburg by the Simplo-Filler-Pen Co. itself in 1919. The *Diplomat Safety Pen* is still known to connoisseurs today as a particularly noteworthy piece. It had a short cap with a forked cylindrical raised section. One moved the nib out with the fork, which was on the back of the barrel. Turned around and screwed ed the rigid writing tube to the flexible nib, the Simplo-Filler-Pen Company also offered a small assortment of "stylos"—ink-writers with stiff writing tubes.

The firm, renamed "Simplo Füllfeder GmbH" in 1914, was able, within just a few years, to develop additional writing instruments that gained world fame then and rank as classic fountain pens today. An outstanding example is the *Meisterstück* (Masterpiece), which has been on the market since 1924 and established the firm's tradition.

In 1919 the proprietors opened their first Montblanc shop in Hamburg. Between 1921 and 1923 others were opened in Berlin, Leipzig, Breslau, Cologne, Hannover and Bremen.

They understood from the start that outstanding quality and exciting advertising are the heart and soul of business. In 1926, four hundred employees in various branches manufactured Simplo products. Branches in Berlin, Barcelona, Lugano, and Copenhagen formed bridgeheads for exporting. Along with writing instruments, they offered loose-leaf business books produced by Simplo Imex. At the beginning of 1919 the North German Vulcan Fiber firm, which supplied the vulcan-fiber and hard-rubber raw material for the manufacture of writing instruments, was absorbed. In 1921 the firm diversified further. The Bruyère Pipe Corporation, under the direction of Christian Laussen, manufactured and distributed high-quality smoking goods. Later, leather goods for writing were also produced from the firm's own facility in Lämmerspiel, near Offenbach. Another writing goods firm, the "Astoria Füllhalterwerk GmbH," was to follow in 1933 to maintain production in the area of reasonably priced writing instruments. The chemical factory of Carl Doutiné, acquired in 1938, handled the production of Montblanc ink.

In 1934, the firm's name was changed again. From then on the writing instrument firm did business as "Montblanc Simplo GmbH." With the success of the Montblanc brand, the variety of its line also grew. The 1927 price list includes Montblanc safety pens, self-fillers (lever-tube system), the *Meisterstück 4810* (safety or push-button-tube system), fireproof pens (push fillers, unbreakable, "fire-tested" materials, no "cheap" celluloid), octagonal pens (safety system), low-priced safety pens for students and young people, pens of precious metal (covered with 14-karat gold or 900/000 silver), ink writers, filler pencils, schoolbook bags, men's note pads, traveling ink bottles and men's pocket notebooks, photograph albums, paste, briefcases and mechanical pencils. The series of writing instruments were available in elegant black or marbled colors and in three, four, or more sizes.

The pride of the house: The Meisterstück (Masterpiece) series of 1936.

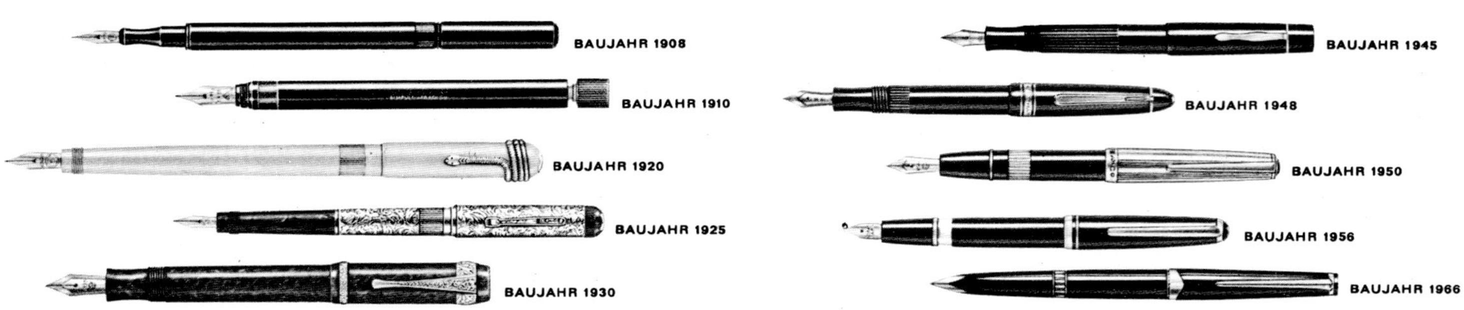

Historic Montblanc writing instruments.

In 1935 the firm was able to patent an automatic pressure pencil, the *PIX*. Filler pencils were also available in giant or midget sizes, and even four-color drawing pencils.

The 1936 price list made it necessary to bring order to this vast system with a numbering system. The first digit of the catalog numbers indicated: 1: Higher price range (Meisterstück), 2: Medium price range, 3: Lower price range. The second digit indicated the system: 0: Safety pen, 2: Self-filler (push filler), 3: Control filler (piston filler). The third digit indicated the nib size:), 1, 2, 3, 4, etc. For example, the number 126 means: 1: Meisterstück, 2: Self-filler, 6: Nib size 6.

It is noteworthy that the Meisterstück class was available as a safety or push filler until the end of the 1940s, with just two exceptions. The piston filler system with its advantageous ink-level window was available only in the medium and lower price ranges (Reflex) as well as in ink-writer and other forms. At the end of the 1930s Montblanc was still offering the glass pen as a reasonably priced writing instrument, in either safety of push-filler form.

During World War II the realm of Montblanc writing instruments was a quiet one. Almost 80% of the firm's production had to be delivered to government offices. Branch firms in France, Spain, Holland and Denmark produced Montblanc writing instruments under license. The Hamburg factory was badly damaged.

When production resumed after the war's end, the *Meisterstück* series was made only with the piston filling system. In 1948 the 146, 144 and 142 series were made in their typical form, which has been retained almost unchanged to this day. In 1955, by which time 750 people were employed again, the rest of the assortment was suited to modern-day taste. The medium-priced 200 series included soft, rounded shapes and offered its first modern design in the *Wing Pen* with a velour tip. Along with reasonably priced standard types, students could also buy the *Monte Rosa School Pen*. In 1952 the round-point *Rollerball* inkholder was presented as an answer to the popular ballpoint pen.

The 1960s brought a complete change of generations and a model policy to match. Slim and stepless, with small wing or intarsia nibs (30s models) were offered in luxurious, medium and reasonably priced varieties. The classic *Meisterstück 149* in diplomatic format was still sold as a special model. The Montblanc production process became more and more modern and practical. The old celluloid had given way by 1960 to modern injection methods and thermoplastic materials. In the 1970s, tastes changed again. The conventional Classic Line introduced in 1968 was followed by the *Carrera* fountain pen for students in 1970. In 1973 the *Noblesse* appeared, a slim cylindrical all-metal line, which was expanded into the *S Line* in 1980.

Meanwhile the *Meisterstück* series with its classic piston mechanism had progressed from an outsider to a big seller, and has long since been regarded as a prestige item as in earlier days. Now as then, Montblanc ranks among the world's leading manufacturers of high-quality writing instruments.

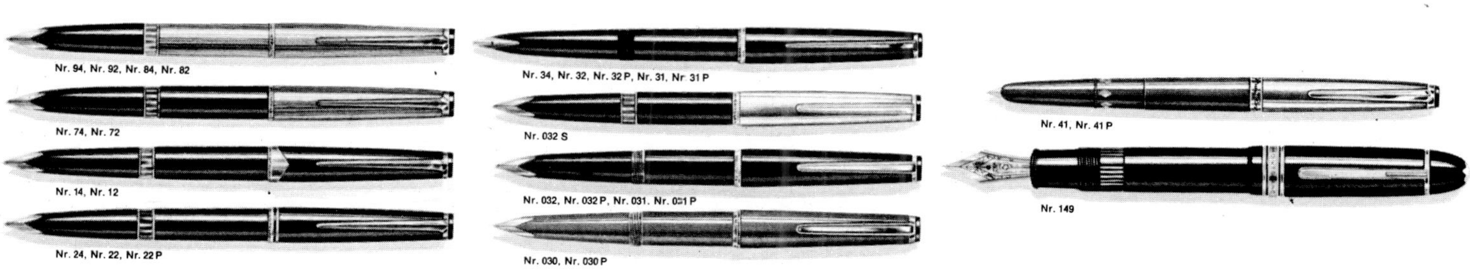

The modern assortment of 1967.

From the Stylo to the Indelible Pen

The fountain pen had scarcely been invented in the 1880s when it was threatened with competition from a half-brother. The Englishman J. D. Carter thought up an ink-writer in 1882 that had, instead of a point, a precious stone on a pointed gilded writing tip. Through the jewel, according to the patent description, a fine channel was bored, through which ink could come out. The special feature of Carter's invention was the ink regulation. A fine long needle in the ink reservoir conducted the ink through two valves to the writing tip. If one pressed on the paper when writing, the valves opened, released the flow of ink, and let air in at the same time.

Despite numerous improvements by Steward and others, the stylograph had a hard time catching on with the public. The new writing instrument lacked the flexibility of the gold nib that allowed an expressive script with fine hairlines and strong heavy strokes. Under too much pressure it tore the paper, dried up, or spattered so heavily that its negative image soon extended to the fountain pen too. Overly cautious people warned others of the stylograph, calling it a "gravedigger" of cultivated handwriting. Only A. T. Cross was able, in 1896, to develop a halfway usable device.

When great German firms such as Montblanc (1910), Kaweco (1912) and others offered ink-fillers with writing tubes of platinum-iridium, the use of the simple, economical device slowly grew. It was used particularly in offices where the modern typewriter had established itself and there was no longer a need for an artistically created, calligraphic handwriting. Entries in a journal or ledger were to be made quickly and legibly by a bookkeeper, often with several carbon copies. And the stylo did the job more easily than the copying pencil previously used. And the office manager saved money. The price for a stylo was barely one-tenth of what had to be spent for a serviceable fountain pen with a gold nib.

There were reasons for the low cost of manufacturing. Until the 1920s the ink stylos were made as pure insertion fillers to be filled with an eyedropper and had, except for the types whose needles could be adjusted, no mechanism. Their malfunctioning, of course, led to the writing tube getting plugged with bits of paper or dried ink. Then too, the laborious filling was burdensome. If the stylo was overfilled, the remaining bits of air pushed the ink out at the wrong time. Only modern designs at the beginning of the 1920s brought improvement. Then, for example, Montblanc offered a self-filler with a push-button that could be filled in seconds. It had a fine tube, rounded at its writing end, made of a platinum alloy, which scarcely wore out, unlike the gold-plated surfaces of earlier writing tubes. Inside it a gold wire moved, weighted with a bit of lead and projecting about a millimeter out of the writing tube. The gold wire cleaned the ink channel when shaken. When at rest, the wire automatically fell forward again and closed the ink reservoir off, creating an airtight seal to contain the ink. When writing, the slightest pressure was enough to push the wire in and let as much ink as was needed for quick, even writing pass through the steel tube. Separate ink and air passages guaranteed undisturbed ink flow.

The first Montblanc stylographs of 1910 had a coiled wire as an ink feeder. They were filled with a pipette, like the insertion filler.

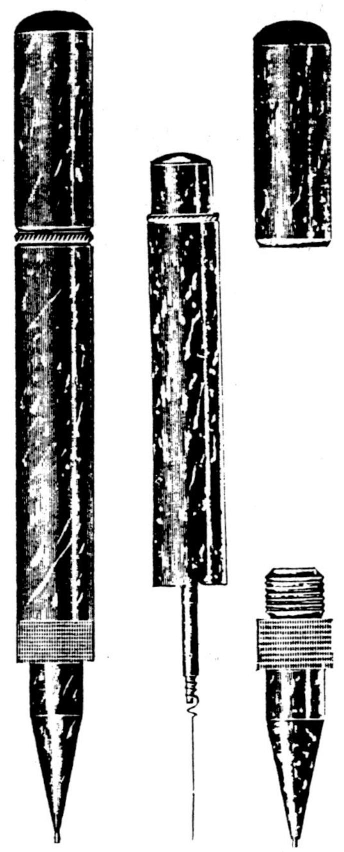

CARTER 1882

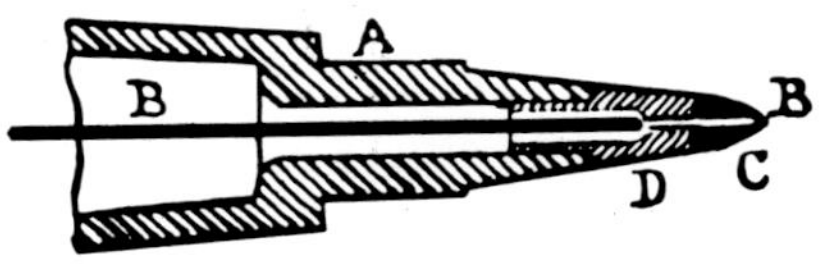

STEWART 1888

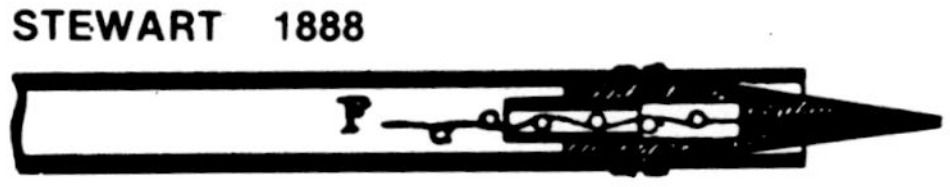

CROSS 1896

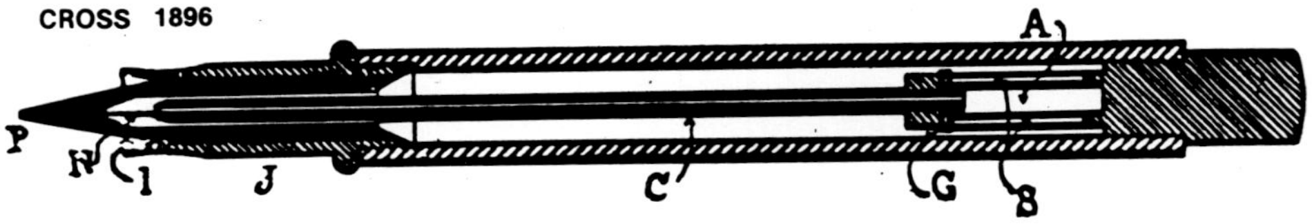

Above and right:
The first ink writers with a cleaning apparatus.

"Millions swear by the red-ringed 'Ink-Kuli.'" Thus Riepe of Hamburg advertised the masses' new writing instrument of the 1930s.

Wilhelm Riepe of Hamburg made the stylograph a writing instrument for the masses in 1928. His ink ball writer became the writing device for everyone. Just as we use the ballpoint pen today, the ink stylo was used then for making quick notes or sketches. Later it was fitted with a piston filling system and an ink-level window. Sketching and drawing could be done splendidly with the ink stylo. Unlike the pen, its strokes were always of the same thickness. This was an advantage that technicians and designers at the drawing board valued. In American drawing offices the unlikely ink stylo was only used when work with the traditional drawing pen or instruments was too impractical.

Riepe made use of the opportunity. Building on the tested ink passage tube with its cleaning needle and falling weight, the firm developed the simple ink stylo into a precision drawing instrument. This was a difficult undertaking, for instead of thin writing ink that consisted of water and dissolved pigments, thick drawing inks with solid particles of soot in shellac solution now flowed through the fine channels of the passages and the writing tube. The difficult aspect of the indelible ink that dried on the paper, impervious to water or erasing, was that it dried much faster than ordinary ink in and on the writing tip too.

The ink ball writer, as the successor to the earlier stylographs, went through numerous technical developments in the 1950s and 1960s. Against the ever stronger competition of the modern ballpoint pen, fineliner, and ink ballpoint, though, it could not survive. The indelible ink filler stylo, put on the market in 1954, revolutionized technical and graphic drawing all the more remarkably around the world. Thanks to its constantly precise stroke width that allows lines to be drawn to tenths of a millimeter, it is used today by builders, architects, and graphic artists all over the world. Whether for advertising placards, buildings, bridges, machines, airplanes, or even spacecraft, the modern indelible filler stylo was involved in their origins.

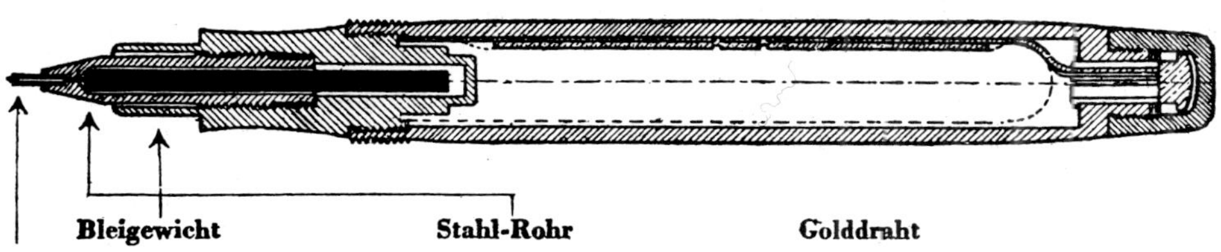

Cross-section of a comfortable Montblanc ink writer of 1936, as a self-filler with pressure mechanism. A moveable gold cord, weighted with lead, provided free ink flow.

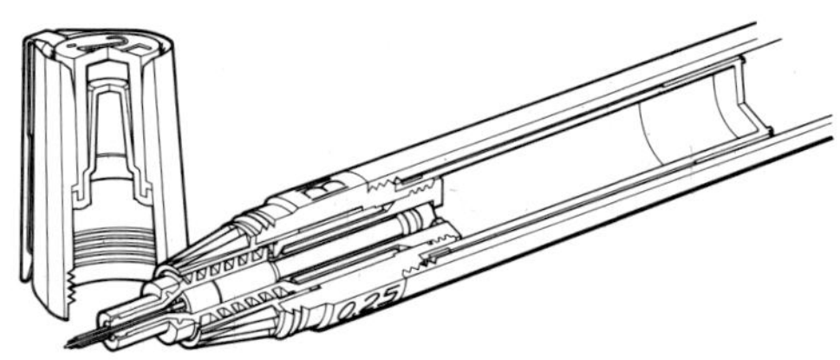

The principle of the ink writer was adopted for the indelible ink fountain pen. A cleaning needle with a lead weight provides undisturbed flow of the pigment-rich, quick-drying drawing ink.

The successor to the former Stylos revolutionized technical drawing: From 1954 on, the indelible ink fountain pen could be used on the drawing board for more precise drawings than could be made with earlier drawing pens and instruments.

A 1955 advertisement for the "Ink-Kuli": Writes like a pencil but with flowing ink.

rotring

The first "Ink-Kuli" ink ball writer, 1928.

The year was 1928. The teletype was invented. Alexander Fleming discovered penicillin. In America, where the twenty-seven year old cartoonist Walt Disney had just given the world the cartoon-film character Mickey Mouse, the Hamburg businessman Wilhelm Riepe, who until now had traded in dried fruit and herbs, was paying a visit. By this time Riepe had often made business trips to America. But the year of 1928 was to lead his career in an unexpected direction. In an American stationery shop he discovered a device that awakened his interest particularly: a fountain pen without a nib. Curious, Riepe bought several samples of the "Stylopen," which had a small writing ball instead of the usual gold or gilded steel nib. The pen was filled with an eyedropper after the barrel was unscrewed.

On his ocean voyage homeward, Riepe thought about how he could improve this instrument, which wrote very badly. For one thing was clear to him, a fountain pen without a nib could be produced and sold more economically that the fountain pen with a gold nib, usually restricted to a class of customers with plenty of buying power. He also knew of doctors and druggists with whom he often dealt in the herb trade, to whom such a writing instrument would be well suited for writing prescriptions, for one could also write carbon copies with the "stylo" because of its stiff writing tube.

After he returned from America, Riepe and his employees began the further development of the American "Stylopen." On October 24, 1928 they succeeded. A fountain pen with a writing tube and point was born and given the striking name of "Tintenkuli"—"Tinte" (ink) for the medium and "Kuli" (coolie) for the term then used for an auxiliary worker. The "ink-coolie" with its non-wearing ball of V4A steel, visible ink level and safety filling pump became a public favorite. One could even test it for a week before buying it.

After Riepe had patented his improved model, his own production of instruments and ink began in Altona in the mid-30s. Meanwhile this pen had been accepted in society. Every good stationer offered a line of them. During World War II the "ink coolie" became every soldier's writing instrument. The Riepe firm, now renamed "TiKu GmbH," was under contract to produce it exclusively for the German Wehrmacht.

After the war ended, the Riepe firm received raw materials from the British occupation forces in return for the delivery of a thousand pens per month. Production and exporting grew slowly toward the end of the 1940s. As it had at the very beginning, a forward-looking idea from America helped out. There technicians had recognized that the use of the "ink-coolie" instead of the time-honored drawing pen would make the work of preparing technical drawings easier. The only problem was that the ink did not have the necessary resistance to erasure, light, and water. Changes in its construction by the Riepe designers made the pen capable of being filled with indelible ink.

In 1953 the first tubular indelible ink stylo was born. Tested at first in the USA, then introduced in 1954 in Germany and other countries, the classic tubular indelible ink pen with the red ring opened a new era in drawing. Since one could work with it must more quickly, neatly and easily, it was given the name *Rapidograph*. The first "Rappi," as it was nicknamed by the technicians and graphic artists, was soon fitted with a more practical replaceable ink reservoir. Gradually the legendary "ink coolie" lost its popularity. The indelible ink stylo, its successor, became all the more popular. With the introduction of the "rotring micronorm" in 1968, precision was stressed to the limit. Now it had become possible to make microfilm-precise drawings with the indelible ink stylo. The firm's technical writing and drawing products were rounded off with stencils, compasses, drawing plates and other products needed in technical drawing.

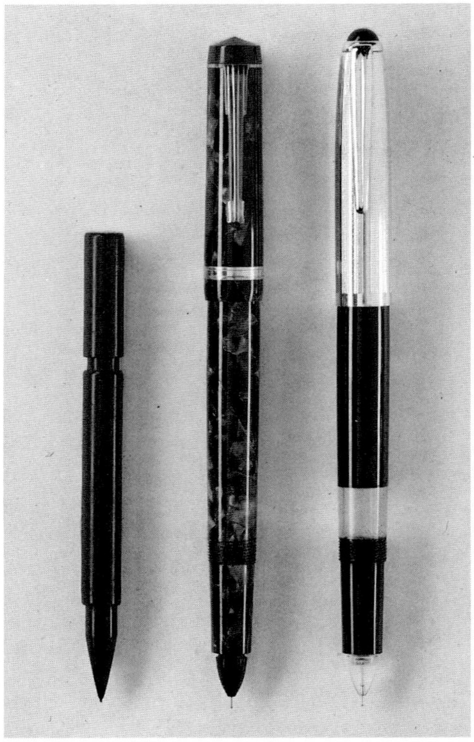

The ink ball writer over the years: from the pipette-filled type of 1928 to the piston-fillers of 1958 and 1978, the latter with a metal cap.

Today "rotring" is one of the most important manufacturers of an inclusive system of drawing instruments. Along with a variety of products for writing, drawing, and designing, the line of products for creative spare time activity keeps growing in importance. Articles for graphic artists and amateur painters were added to the line in 1983. By acquiring the Aristo firm, another step was taken toward a successful future. Along with the manufacture of classic products, it was possible to put a suitable computer system for future use in design and technical drawing on the market at the right time. Despite the growing future development of "rotring" products, the indelible pen is still the firm's most important seller inside and outside Germany.

A hand-painted display advertised the "best friend" in shop windows.

In Altona, a suburb of Hamburg, ink-kuli production began in the mid-30s.

Vendors dressed in Chinese coolies advertised the ink kuli on the streets of Hamburg in the 1950s.

A snapshot of a rotring advertising poster of 1952.

Ball Point Pen

The Fountain Pen—

"The pen is mightier than the sword."
from Eurydice

The American writing instruments returned to Germany with the American soldiers. The Parker 51, first made in 1940, was often imitated. A 1948 advertisement.

Fountain pen production resumed only sluggishly after World War II. During the war most of the European and American manufacturers had been compelled to reduce their production of writing instruments or even give it up entirely in favor of armaments or other goods that were needed in the war. Important precious metals needed to make pens, such as gold, silver, osmium, and iridium, as well as stainless steel, were available only on a contingency basis or not at all. In Europe during the 1940s, gold pens ranked as desirable rarities; chromium-nickel pens dominated the small supplies. The further development of the fountain pen came to a worldwide standstill. And just at this time the ballpoint pen, the king of writing instruments' toughest competitor, was born. The idea of writing with a ball instead of a pen point had already kept many inventors busy. Without success, of course. The breakthrough, curiously enough, was achieved independently by two men at the same time and in the same way.

The Hungarian Lászlo Biró and his brother Georg were able to escape the dangers of war in Europe and flee to South America at the beginning of the 1940s. The most important item in their "luggage," so to speak, was the idea that they developed in the New World: making a ball in a metal fitting filled with ink roll over paper to draw a continual, sharply defined line. In Argentina they found an interested partner, the firm of Eterpen S.A. Thanks to international connections, the new pen could be manufactured under license in England in 1944, and by their North American contract partners Faber and Wahl-Eversharp in 1945.

At almost the same time, an outsider turned up. In the winter of 1945 the industrialist Milton Reynolds contested the Biró brothers' first right to the invention. His *Satin-Flo* proved to be a competitor from the start, bringing on a tough competition among the many manufacturers of ballpoint pens. They tried to beat each other in pricing ($12.50) and in advertising.

The original inspiration for the new pen soon gave way to a deep disappointment. Neither Eversharp and Faber, nor the outsider Reynolds could keep their promises of quality. The ballpoint pen proved to be an undeveloped, unreliable instrument, similar to the very early fountain pens. It spattered, smeared, and seldom wrote immediately. After a few years Reynolds, one of the inventors of the ballpoint pen, had to give up and disappeared from the market. Eversharp's fate was scarcely better. The high investments in production facilities to meet the originally fast-rising demand, higher manufacturing costs that had originally been expected, a murderous price war, and the costly recall of thousands of wrongly made pens brought Eversharp to the brink of ruin.

Sheaffer had a bit more luck: The firm took more time and began ballpoint production only in 1947, with the *Stratowriter*. Like Parker, which introduced the famous *Jotter* only nine years after Biró and Reynolds, Sheaffer maintained a full line of fountain pens. In 1949 the unusually slim *Thin Model* series appeared, which could be filled at a single stroke by pulling out its piston ("Touch Down"). In 1952 came one of the most curious of all filling systems: the *Snorkel Pen*. Rather like the periscope system of the submarine, a tube could be extended under the nib and dipped into the ink bottle. In this elegant way the Snorkel pen avoided the otherwise unavoidable discoloring of the nib, the forward section, or even the fingers with ink. In 1959 Sheaffer created its famous oversize pen, inlaid similarly to intarsia work in the plastic of the forward section. A Sheaffer pen can be recognized at first glance by this feature to this day.

In Germany too, the ball point pen became popular quickly. A 1949 Reynolds advertisement.

The great English brands of Onoto and Swan watched the futuristic designs of Parker, Eversharp, and Sheaffer very closely, for they were creating a sensation across the Atlantic in the British Isles as well as in America. Hesitantly they

Decline and Rise

designed their models after the great American examples. Montblanc stuck to the tried and true, carrying on the tradition of the *Meisterstück*. The 142, 144 and 146 model series offered a unified streamlined design in a variety of sizes and diameters in 1948. The screw mechanism of the piston was at first made of metal and guaranteed decades of smooth, dependable functioning. The glory of the oversize "Diplomat 149" model of 1952 was the big gold nib with rhodium decor.

The Pelikan firm likewise trusted in the classic tastes of European customers. In 1950 they presented the successor to the Series 100 that had sold successfully for more than twenty years: the 400 line with its typical dark green or brown tortoiseshell barrel band. With this Pelikan scored another international success. All models bore a big 14-karat gold nib with the special "Bandzugfeder" (band pen) cut that produced writing that looked like calligraphy. For the first time the clip took the form of a pelican's bill, which would become famous later. The 500 and 600 models had their cap and filling grip covered in 14-karat gold, and the barrel of the 700 model was plated in gold. In 1955 Pelikan began to produce ballpoint pens.

Other well-known German names, such as Soennecken, Kaweco, Osmia, Faber-Castell, Matador, Merz & Krell, Staedtler, Greif, and Artos likewise tried to duplicate their pre-war success with new models. The office-supply manufacturer, Geha, in Hannover began to make fountain pens with reserve tanks. A clever apparatus under the nib allowed the writing of a few more lines when the main tank ran dry by pushing a lever. The glue manufacturer, Uhu, tried to diversify its program with a line of fountain pens.

A humorous advertisement for the everyday ball point pen in 1956.

Meanwhile the American ballpoint pen was conquering the world. Among its best promoters were the American soldiers who proudly introduced the new writing instrument to their allies and opponents as a symbol of American progress, despite its originally high price of 50 D-Mark and more. The traditional fountain pen with its big sensitive nib and laborious filling from an open ink bottle, as well as the good old mechanical pencils of various types, with their ever-breaking leads had to step back because of their weaknesses and let the ballpoint pen have star billing. For the newcomer had left its formative years behind it and had risen to the status of a reliable everyday utensil for almost every purpose.

Production of the ballpoint pen, at first so difficult, no longer called for special operations. Almost any halfway prosperous small business could afford the needed machines and start production at low cost. The ballpoint pen, once expensive, was now produced by the billions all over the world and descended to the level of a giveaway article that cut-rate dealers distributed at fairs and markets along with a dozen razor blades.

Still in all, as practical as the rolling ball was for everyday service, this modern writing tool could not contribute to the development of a cultivated individual handwriting. The writing teachers complained most about it as they observed the growing deterioration of handwriting in the schools. They soon gave the ballpoint pen bad grades as an instrument for writing instruction. "Back to the fountain pen," was their demand, for it alone could develop the untrained child's hand into neat, disciplined handwriting.

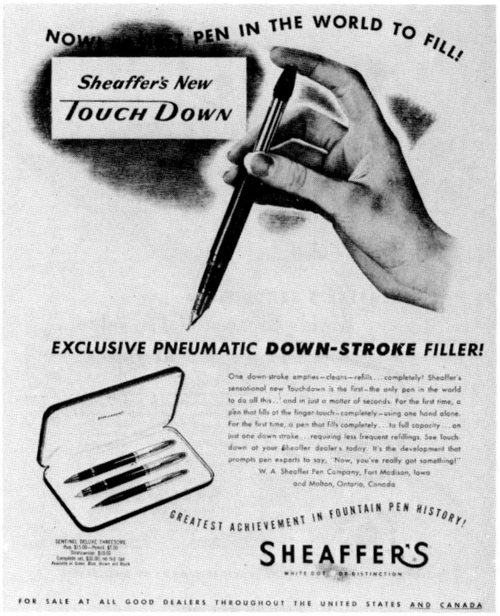

The new "Touch Down" filling system: Empty, clean, fill—all with one touch of a finger. A 1949 advertisement.

The salvation of the fountain pen, at least for the time being, came from America, as so often before. Waterman and Sheaffer remembered the invention of the Frenchman Perand, who had patented small glass cartridges for refilling Jif-Waterman fountain pens in 1927. The sensitive glass tubes were replaced by plastic cartridges. In 1953 Waterman presented the "CF," and in 1955 Sheaffer brought out its own low-priced ink-cartridge pen. This gave a signal that the German manufacturers, loyal to the fountain pen, seemed to have been waiting for.

Diplomat was the first to utilize the American filling cartridge. It made filling the pen with ink as clean and trouble-free as changing a ballpoint cartridge. In 1960 Pelikan followed with a fully new school fountain pen developed especially for children, the "Pelikano," and it was not long before Geha replied with a similar type. They helped the modern cartridge filling system become established in Germany at last. The inkwells in the schools became superfluous overnight. Every pen had a spare cartridge for fast replacement. Ink bottles that spilled or got broken in schoolbags were a thing of the past. For years Pelikan could scarcely keep up with the demand for the new cartridge pens, which were the result of long customer research in many schools. Orders from dealers had to be cut drastically and the wares divided among them. Many dealers advertised in their home-town newspapers: "Pelikano cartridge pens—just in stock . . ."

The glue manufacturer Uhu also sold fountain pens and ink in 1949.

Other German and European manufacturers followed the trend to the ink cartridge. The cartridge fountain pen, especially as a school pen, gained great popularity in Germany in just a few years. The higher-priced fountain pen for adults suffered in the process. It moved more and more into a shadowy existence at the edge of the product lines of the great German writing instrument manufacturers, pushed out also by new instruments such as fiber- and felt-tip pens, which were crowding into European and American markets from Japan.

Along with Montblanc in Germany, the lordly fountain pen owes its further development particularly to Parker in America. In 1956 the Parker 61, the first fountain pen without moving parts in its filling system, appeared. A reservoir, composed of spheres of different sizes, filled by capillary action when dipped in an ink bottle. Jif-Waterman brought out a model with a similar system at the same time as Parker. In this case a felt-like reservoir material similar to that of a cigarette filter was used.

The Parker 75, first issued in 1964 for the firm's 75th anniversary, still ranks among the classics of fountain pen history. Just like its predecessor, the VP (Very Personal), it has an adjustable nib that can be set at the user's preferred writing angle. A special series of 4821 pieces came out as the "Spanish Treasure." For the covering of sterling silver, melted coins from a Spanish galleon that had sunk off the coast of Florida in 1715 and been salvaged in 1965 were used.

In the 1970s all was quiet in the world of the classic adult fountain pen. New mass-produced writing instruments like fineliners, ink rollers and fine-line pencils won high places as everyday tools for everybody. The typewriter moved into the private sector, the photocopier, computer, and printer dominated modern offices.

A Staedtler advertisement.

The "Senator" piston fountain pen, circa 1950.

To the extent that adults still used it at all, the fountain pen was only hauled out now and then. It was used for an impressive signature on an important letter or contract, for a personal resume for a job application, or for a love letter to a much-desired partner when the use of the graceful pen appeared to be unavoidable. In America, the land where the fountain pen had been invented and highly developed, writing with that pen was almost forgotten.

Many small manufacturers who were not able to convert to the production of school pens or other types either gave up or concentrated on other types of business. The names of Kaweco, Soennecken and Brause in Germany, Onoto and Swan in England are examples of what happened to dozens of others. Other firms tried to draw customers' attention to themselves with new, unusual designs, going against the trend of the time. A rebirth of the Bauhaus architectural style of the 1920s particularly inspired people with a well-developed feeling for form.

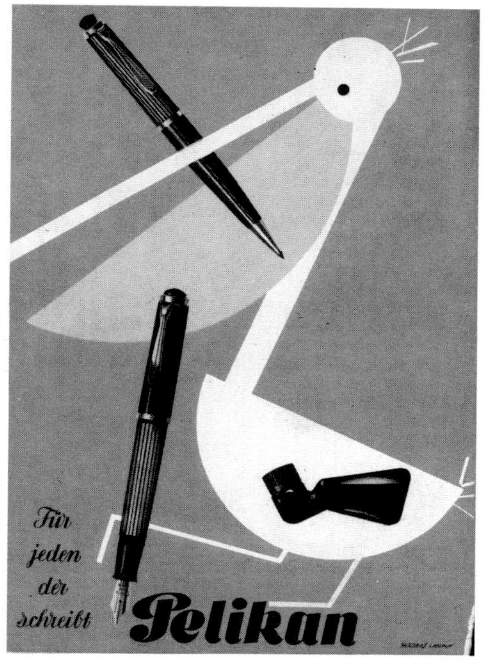

The black and green Pelikan in new form. A 1952 advertisement.

The traditional family business of C. Josef Lamy GmbH in Heidelberg surprised the world of writing instruments in 1966 with an extraordinary fountain pen design, the *Lamy 2000.* This undecorated, understated series of piston fillers, ballpoint pens, mechanical pencils and four-color pens consisted of striped black plastic combined with brushed stainless steel and a spring clip cut out of solid stainless steel. It took some time for this design, courageous for its time, to catch on with the public. After more than twenty years, the *Lamy 2000* is regarded as a classic by today's connoisseurs. Other pens like the *Design 20, cp 1,* and *Unic* series followed in the same line of development, which was raised to the level of a business philosophy by Lamy of Heidelberg.

But even these successful attempts to supply the public with technically and artistically highly developed, valuable writing instruments did not change the trend of what people were buying. The time when one proudly let the golden clip of an image-boosting fountain pen sparkle on one's breast pocket was gone. Low-priced, reliable writing instruments for the masses, above all, the ballpoint pen, did their job too. An average of about 250 million ballpoint pens were produced per year in Germany alone. If their strokes were too ordinary for someone, he could always turn to the fiber-tip pens. Their stroke, 0.2 mm wide, could be mistaken for that of a traditional pen. Scratchiness when one began to write, a writing tip that grew wider all the time and finally became dull and scratchy when the ink supply in the reservoir ran out or dried up prematurely were outbalanced by the advantages: no spattering, no refilling from an open bottle, and readiness for writing or drawing at any time. And finally, the advantages of fiber-tips and ballpoints were united in one other device: the ink ballpoint pen.

Many pessimistic predictions were made. In the year 2000, handwriting would be a thing of the past. The paperless office and the network-linked household with its computer and monitor would be everywhere. Only a few suspected that the classic fountain pen, with its big open nib, had what it took to experience a splendid rebirth.

The forerunners of this development were two big American firms. Parker thought about its own history in 1972. The Big Red, a stylistic throwback to the famous *Duofold* of the "Roaring Twenties," was put on the market as a ballpoint pen and fineliner with a replaceable cartridge. At the same time, it served as an answer to Sheaffer's successful *No Nonsense* pen, a low-priced remake of Sheaffer pens of the 1915-1918 era.

Both series sold by the millions. And the customers were not only people who fondly remembered "historic" writing instruments, but also younger people who regarded these types as something completely new. Both types fulfilled these unconscious longings only incompletely. In a world of plastics and computers, throwaway articles, excess, and destruction of nature in the name of progress, the desire was born in Western civilization to return to genuine values. People looked for means of self-expression in their creative activities. The change in values involved hobbies such as pottery making, spinning, weaving, painting, music, and trying alternative life styles. People wished more and more for products that were

Pelikan has used the ink cartridge since 1960.

worth something and lasted a lifetime. Whatever the subject, people wanted to get away from the uniformity and indifference of the consumer society. Instead of using a typewriter for private correspondence, people turned back to handwriting. And a personal, expressive handwriting can only be achieved with fountain pen and ink. The traditional fountain pen slowly but surely returned to a position of public interest in the early 1980s.

In America, Sheaffer reproduced the *Nostalgic Pen,* a new edition of a favorite type of fountain pen first made in Victorian style around the turn of the century. Omas of Italy presented its series of twelve-sided writing instruments that recalled the successful *Lucens* series of the 1930s. Montblanc sense a growing interest in "the art of writing" and replied with the classic and by now overworked *Meisterstück* series. Pelikan aimed straight at the heart of the consumer whose values had changed, reissuing its famous dark green fountain pens of the 1950s with the advertising slogan: "Timelessly beautiful and valuable."

In the latter half of the 1980s, the king of writing instruments could once more appear in all its former splendor. In the Federal Republic of Germany and many other European countries today, the fountain pen ranks first in sales of writing instruments and creates an undisputed image of prestige and quality.

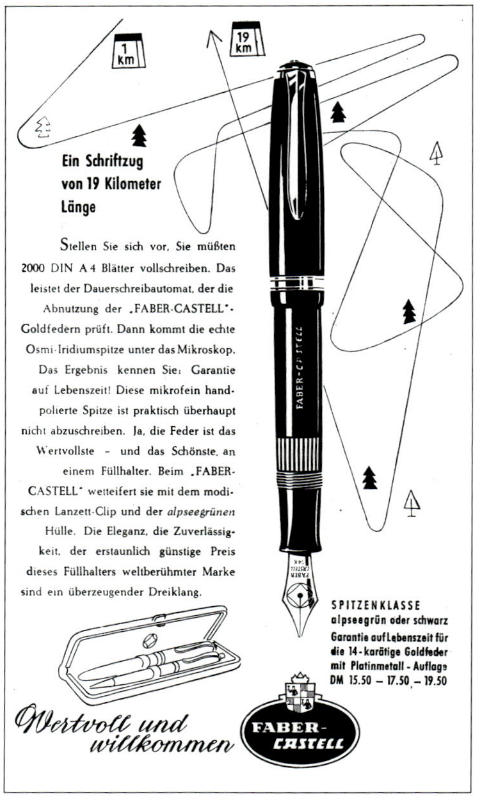

A prestigious Faber-Castell piston filler of 1955.

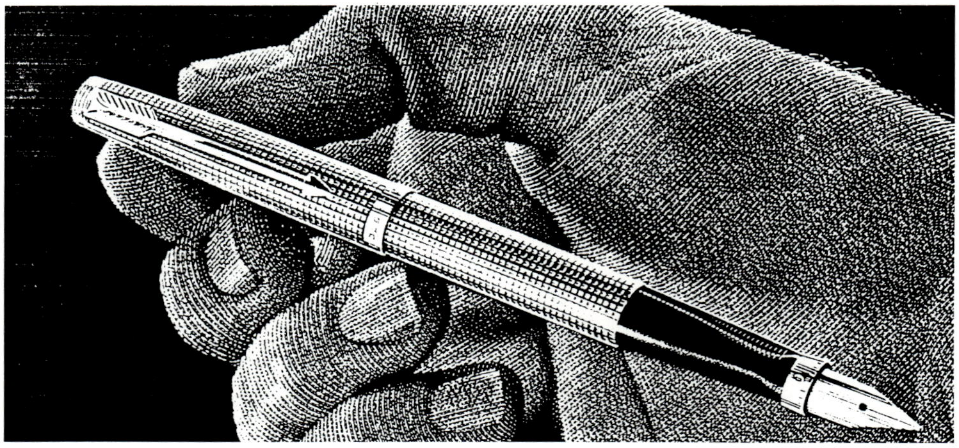

The sterling silver Parker 75 special series.

Eye-filling. The Parker 61 of 1956 had a filling system that used no mechanical assistance.

The Montblanc Meisterstück in the style of the 1960s.

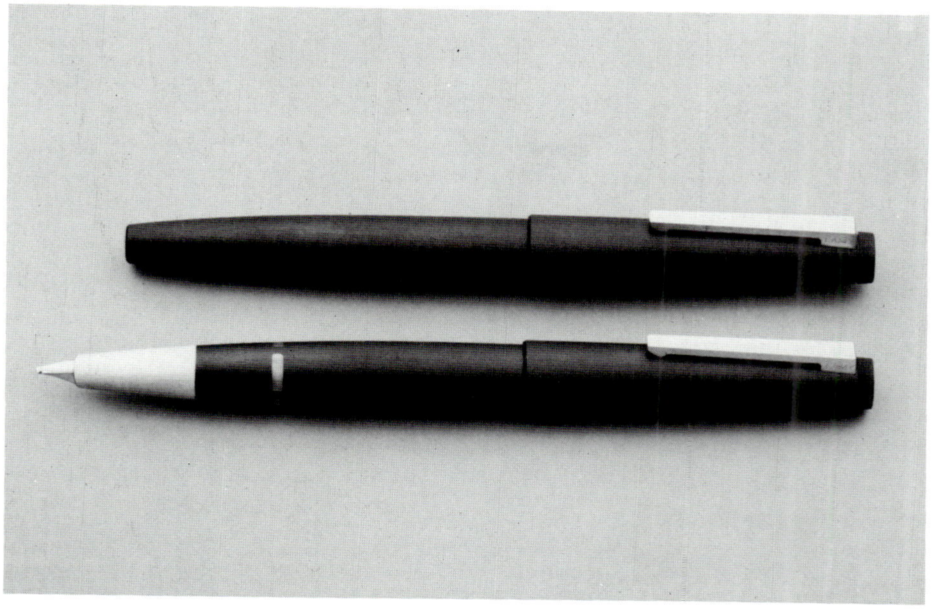

The courageous design of the Lamy 2000 caught on only slowly in the mid-60s. Today this pen is already a modern-day classic.

The Montblanc Meisterstück 149 embodies the "art of writing" again today.

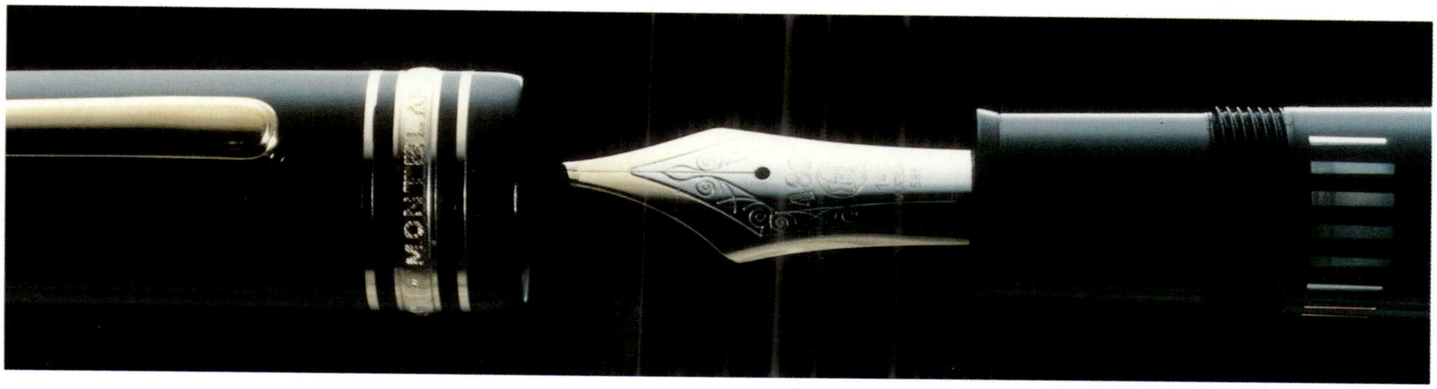

Harmony and precision—the Pelikan Souverän (Sovereign).

The Mobile Advertisement

"Advertising is everything," the inventor of automobile mass production, the American Henry Ford, is said to have told visitors who asked him the secret of his success.

The German manufacturers of writing instruments in the 1920s and 1930s must have had something similar in mind. They translated Ford's viewpoint into "Attention is everything." The flood of typical advertisements in periodicals must not have resulted in resounding success. The growing mobility in those days, thanks to mass and individual transit, quickly gave the advertising professionals the idea of using the progressive image of the automobile and airplane for their own purposes. High above the heads of the crowd, the cars and the housetops, the quality and prestige of the advertised writing instruments was raised before the eyes of the consumers. Whether the mobile advertisement had the desired effect can only be guessed at. At least it was original!

In the 1920s the airplane was already an exclusive means of transportation.

The famous Montblanc truck fleet in downtown Hamburg.

Around the turn of the century, balloonists drew a crowd.

During overland sales trips, the advertising pioneers had to deal with flat tires.

"That takes care of Christmas business!" Pelikan proudly proclaimed to its astonished dealers in 1931.

Lamy

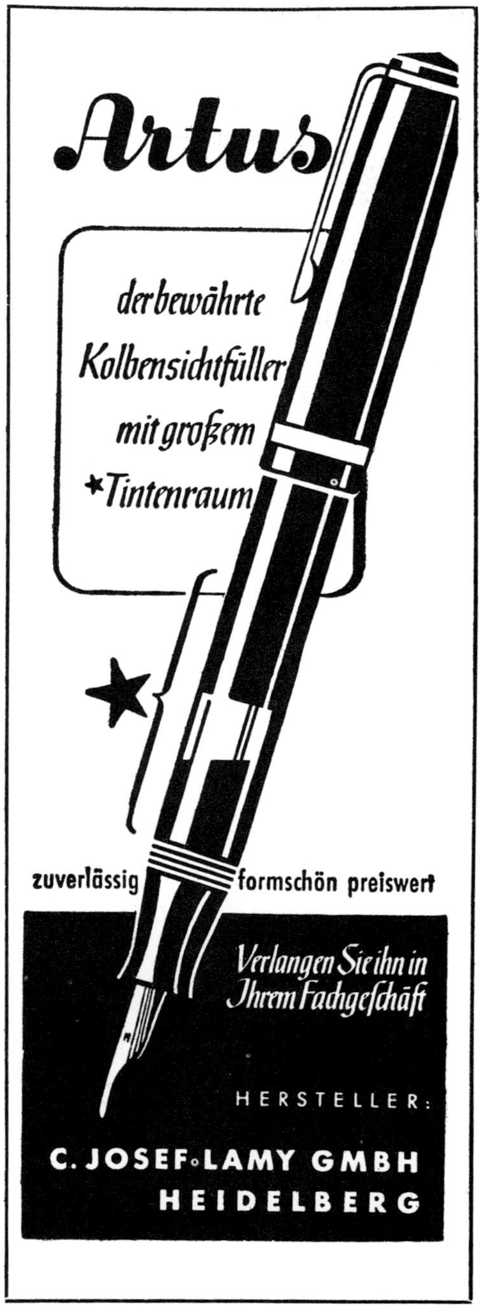

The Artus piston filler of 1949.

Two events shaped the history of the firm of Lamy: the more recent lies somewhat more than twenty years in the past. In a day when writing and calculating machines made by Olivetti of Italy, cars made by Citroen of France and electronic and audio equipment by Braun were the pioneers of progressive merchandising, a new, unusual writing instrument took its place in the ranks of the increasingly numerous "Bauhaus Neoclassics:" the *Lamy 2000.*

The unadorned and compact fountain pen encountered a blend of excited acceptance and hesitant skepticism from the start. The massive clip was too clumsy, the harmonious blend of stainless steel and matt brushed black plastic was too daring. Nobody at Lamy expected this new kind of writing instrument design to be accepted at once. But they were convinced that the *Lamy 2000* would usher in a new era in the firm's history.

The second event began with the decision of C. Josef Lamy, born in Furtwangen in 1898, to spend his apprentice and journeyman years in America. At the Janeville facility of the giant Parker firm, the young German got acquainted with the production and marketing of writing instruments in the New World. In 1929 Lamy returned to Germany to put his American experiences to good use as a director in the German branch, the "Parker-Osmia Fountain Pen Factory" in Heidelberg.

Lamy did not stay long with the firm that had trained him. In 1930 he decided to stand on his own feet. In Heidelberg he opened a small workshop and produced the first Lamy fountain pens, which he sold to acquaintances. The first Lamy products offered nothing remarkable. The lapis-blue *Orthos* pressure filler, with its ink protector, was somewhat reminiscent of the Parker *Duofold* in form and equipment. At the beginning of the 1940s, C. Josef Lamy acquired the Artus-Füller-Ges. Kaufmann & Co. of Heidelberg. In 1943 he put 25 to 30 employees to work making *Orthos* fountain pens and pencils. After the war the fountain pen brand name was changed; the name of *Artus* was used for a reasonably priced school fountain pen in 1949 and heavily advertised.

The year of 1952 brought further innovations. The *Lamy 27* linked the product and its manufacturer's name for the first time. At the same time, the *Lamy 27* was one of the first modern German fountain pen designs with a covered nib. Its ink-passage system, the "Tintomatic," had 21 equalization chambers and is still in use today.

The success of the *Lamy 27* brought a significant upswing to the small firm. In 1955 the number of employees rose to 120 workers and officials, 40% of production went into exports. Subsequent *Lamy 27* models with piston mechanisms, such as the *Lamy 30, 31, 32, 36* and *37,* as well as the modern *16, 17, 46* and *47* cartridge fillers and comparable mechanical pencils and ballpoint pens followed at the end of the 1950s. The Lamy large-volume ballpoint cartridge with its rust-free stainless steel point, introduced in the 1960s, ranks among Lamy's successful products.

The founder's son, Dr. Manfred Lamy, is directing the firm's success back to a significant "design language of functionalism." And consequently, the line that began with the *Lamy 2000* was retained. Big-name designers not only developed the functionally formed product but also its environment: packaging, displays, and trade-fair exhibits. The "Studio" line for young people was followed in 1974 by the *Lamy cp 1,* a slim cylindrical cartridge filler, made completely of metal, that finally made possible, along with ballpoint pens, mechanical pencils, fineliners, and four-color ballpoints, the attainment of the firm's goal of breaking into the market for formally and technically impressive writing instruments. The *Lamy Safari* and *Lamy Alpin* of 1980 presented "good form in Youth Style."

In 1982, a new manufacturing generation expanded Lamy's original design concept. In addition to the instrument surfaces finished, as before,

exclusively in black or silver, a third color, white, was added. The customers were not only a stratum of people aware of form and color, but grew to a larger circle of consumers, won by the Lamy "white pen."

Almost twenty years after the creation of that modern classic, the *Lamy 2000*, its designer, Gerd A. Müller, created a new, technically and aesthetically outstanding "beautiful piece of technology," the *Lamy Unic*.

Today Lamy ranks among the leading producers of writing instruments whose form is determined by their function, true to the philosophy of the world-famous founder of the Bauhaus, Walter Gropius. He once said that it paid in the end to give consideration not only to technical perfection and value for money but also to the artistic worth of the product and utilize it to bring taste and status to the people.

Available in Germany and Europe in almost every type of design: Lamy 2000. A modern classic that influenced and characterized the whole business atmosphere of the house of Lamy.

The typical Lamy advertising since 1975. Shown are the Lamy 2000, the cp 1 and the Safari.

A streamlined Lamy 27 of 1955.

Geha

World War I was coming to its end. In Hannover as elsewhere, signs of a collapse were visible. In the uncertain atmosphere of those years, the twenty-five year old industrial businessman Heinrich Hartmann was wondering about his future. The trained office supplies and stationery salesman decided to go into business along with his brother Conrad, despite all difficulties. On April 1, 1989 they founded the "Hartmann Brothers" firm, or "Geha" for short.

They were not yet thinking of producing goods. They dealt in office supplies such as carbon paper and typewriter ribbon for shipping businesses. The necessary capital consisted of their own savings and some help from their parents. With courage and self-confidence, the brothers stayed in business despite much competition. The Geha trade mark quickly became known inside and outside Germany. The business began to flourish. The number of employees grew to thirty in 1920. Carefully and in modest quantities, production of Geha stenographic notebooks and note pads began. The ambitious Hartmann brothers had ideas, including one for advertising that was unusual at that time. After long negotiations with the postal service, a long-term contract to print advertising on postal banking forms was signed. For five years these advertisements reached 800,000 people per day, and helped the name of Geha establish itself.

In 1922, a long-standing idea became reality. Although the two founders had never worked in a manufacturing firm, they were able to do what no one would have believed four years before and produce chemical office supplies in a small suburb of Hannover. Carbon paper, typewriter ribbons, rubber stamps, glue, erasers, and duplicating materials were the basis of the line. In 1931, the program was extended to include "Geha-Rotary" office machines.

In 1940, the Geha firm already employed six hundred workers. Instead of office supplies, they had to make shell casings, lead bullets, and aircraft compasses. In 1945, the Geha plant was occupied and made the seat of the British military government for Lower Saxony. The firm managed to keep

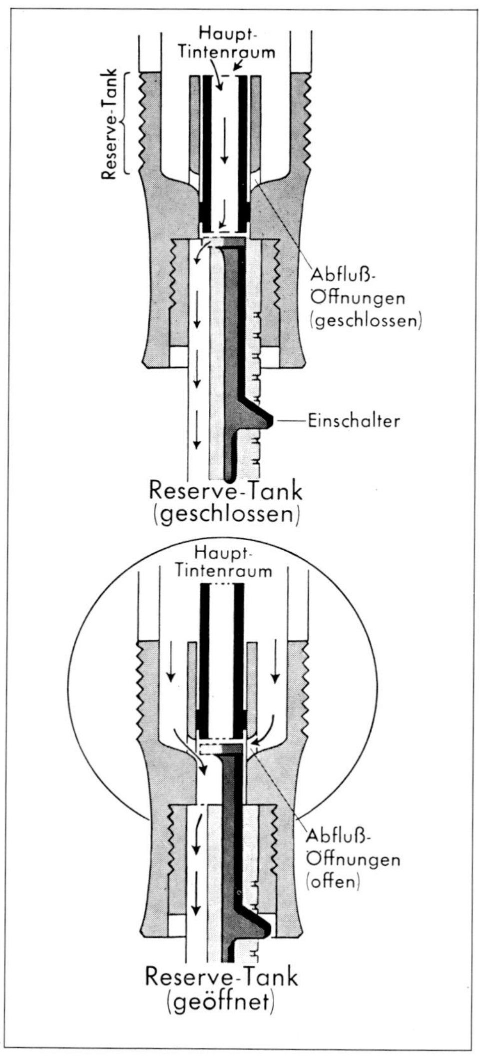

How the Geha reserve-tank piston filler works.

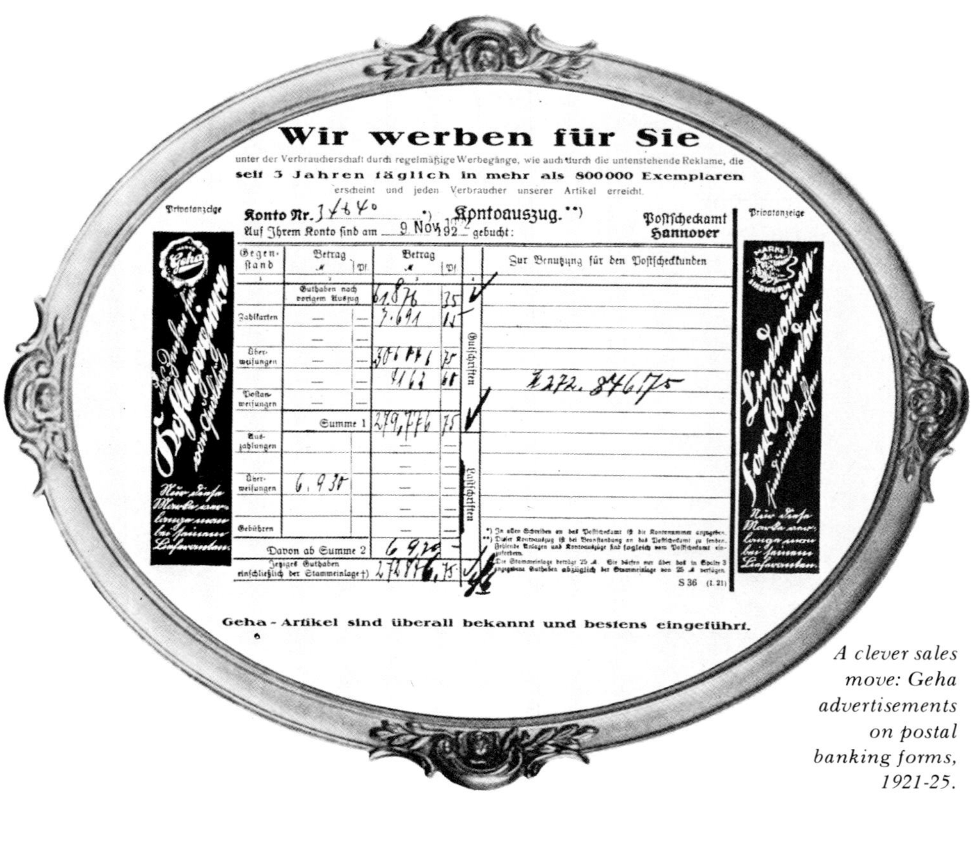

Wir werben für Sie

unter der Verbraucherschaft durch regelmäßige Werbegänge, wie auch durch die untenstehende Reklame, die seit 3 Jahren täglich in mehr als 800000 Exemplaren erscheint und jeden Verbraucher unserer Artikel erreicht.

Privatanzeige

Konto Nr. 14640 *) Kontoauszug. **)

Auf Ihrem Konto sind am 9 Nov 192_ gebucht:

Postscheckamt Hannover

Gegenstand	Betrag ℳ	Pf	Betrag ℳ	Pf		Zur Benutzung für den Postscheckkunden
	Guthaben nach vorigem Auszug		61.876	35		
Zahlkarten	—	—	7.691	15		
Summe 1			279.776	75		
Davon ab Summe 2			6.930			
Jetziges Guthaben einschließlich der Stammeinlage			272.846	75		

Geha-Artikel sind überall bekannt und bestens eingeführt.

A clever sales move: Geha advertisements on postal banking forms, 1921-25.

working in nearby buildings with about 240 employees. In 1949 it was possible to regain full use of all production facilities. In 1950 Geha plunged happily into a new manufacturing program: Geha writing instruments.

From the day they first appeared, the new Geha fountain pens with piston filling and differential operation assured themselves of a special status. The elegant, impressive objects were equipped with a handy reserve tank. If the ink ran out in the middle of a sentence, one only needed to press a small level under the ink passages to be able to continue writing for several more pages. Ranking among Geha's most beautiful and impressive writing instruments from the beginning was an oversize model, the *Geha Regent 720,* with a two-tone nib and the Geha trade mark on the cap ring. Along with this status model (produced until 1956), the Geha line consisted of medium-price (Standard) and low-cost (Student) pens in black, black and pearl gray, pearl green or pearl brown. The *Regent 728* and *Standard 768* were also available as of 1952 with a cap of rolled gold plating and a brown barrel. The *Geha 700* school pen was to be developed in the following decades into one of the most successful school pens in Germany.

The writing instrument program was modernized over the years. At the beginning of the 1960s the classic fountain pen was followed by the elegant *Goldschwinge.* Geha was able to combine the slim form of the modern fountain pen with a big gold nib, solidly anchored in the front section of the pen. This model was still available with piston filling. Following the general trend, Geha concentrated more and more on the modern cartridge filling system. The Geha school pens with reserve tanks, models 701, 703, 707, 708, 3V and many others followed. Geha fiber pens with slim points have become reliable utensils for the daily use of millions of school children.

Today the Hannover firm ranks among the World's leading manufacturers of chemical office supplies, office machines, and writing instruments for the school and the office.

The elegant Geha Goldschwinge of 1961.

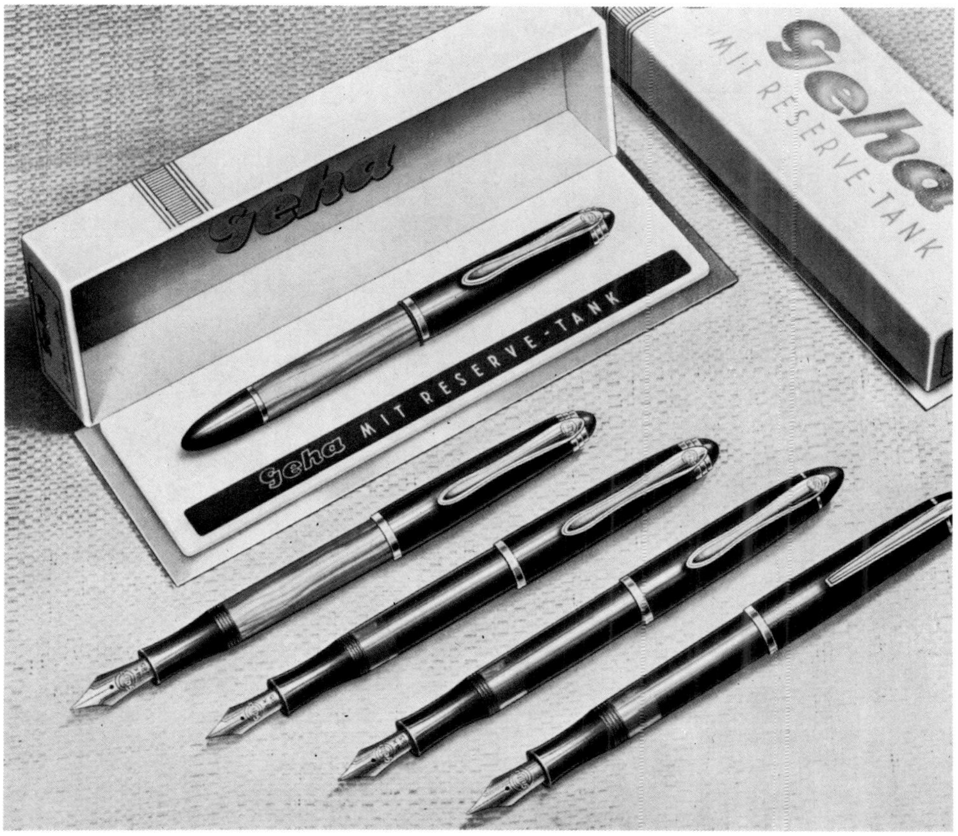

Geha fountain pens with reserve tanks, 1952.

Historic Fountain Pens

The following photos show a selection of the best-known and best-looking historic fountain pens, ballpoint pens and mechanical pencils. The listed values to collectors are approximate guidelines without guarantee. They are based on supply and demand on the international collectors' market. Some pens are extraordinarily rare, and their prices are correspondingly high. Particular brands bring higher priceso plays an important role. Buying privately is less costly than buying at flea markets or from enthusiastic collectors. As a rule, the highest prices are charged by antique dealers and professional dealers of antique writing instruments. Among the most desirable collectors' items are:

Great Britain:
Swan (Mabie, Todd & Co.), Thomas de la Rue (Onoto), Conway Stewart.

Germany:
Kaweco, Montblanc, Osmia, Pelikan, Soennecken.

Italy:
Omas, Aurora.

USA:
Conklin, John Holland, Moore, Parker, Sheaffer, Wahl-Eversharp, Waterman
A number of brands not listed here have a particular value for specialist collectors. It is not so much the age as the condition of each individual pen that matters most in determining its value.

Very Good Condition:
The pen consists of original parts, has not been used, or is not visibly worn (100% of the stated value).

Good Condition:
The pen consists of original parts, which can also have been replaced by expert repairs. Little wear or discoloration (50% of the stated value).
Bad Condition
Pen does not function reliably. Parts are damaged or missing. Heavily scratched on barrel or cap (25% of the stated value).

The Early Years (1900 to 1920)

1. Waterman *Regular No. 14*, insertion filler, 1900. Value: ca. DM 80.
2. Parker *Lucky Curve*, insertion filler, 1900. Value: ca. DM 280.
3. Waterman *Regular 15*, insertion filler, 1904. Value: ca. DM 400.
4. Sanford & Bennet, insertion filler, 1905. Value: ca. DM 80.
5. Doldge *The Spiral Pen*, insertion filler, 1905. Value: ca. DM 40.
6. Waterman *Ideal*, insertion filler, 1905. Value: ca. DM 50.
7. Conklin *Crescent*, tube self-filler, 1908. Value: ca. DM 100.
8. Waterman 20 (oversize), insertion filler, 1910. Value: ca. DM 500.
9. Swan, lever filler, 1916. Value: ca. DM 40.
10. Montblanc No. 2, safety filler, 1915. Value: ca. DM 150.
11. Montblanc pressure pencil, 1915. Value: ca. DM 80.

Pictures on opposite page

12. Waterman 52, lever filler, 1915. Value:ca. DM 80.
13. Waterman 452, lever filler, 1920. Value: ca. DM 300.
14. Whytwarth *Jumbo*, safety filler, sterling silver, 1920. Value: ca. DM 150
15. Conklin, lever filler, 1920. Value: ca. DM 50.

Pictures at lower left

The Golden Years (1920 to 1945)

16. Waterman 452, lever filler, sterling silver, 1920. Value'' ca. DM 150.
17. Blue Ribbon, lever filler, 1920. Value: ca. DM 150.
18. Parker *Duofold Lucky Curve*, pressure filler, 1921. Value: ca. DM 150.
19. Sheaffer *T 8 C*, lever filler, 1921. Value: ca. DM 150.
20. Conklin *Endura*, lever filler, 1922. Value: ca. DM 220.
21. Swan, lever filler, 1922. Value: ca. DM 220.

Pictures at lower right

22. Wahl-Eversharp, lever filler, 1922. Value: ca. DM 120.
23. *Parker Duofold Deluxe*, pressure filler, 1925. Value: ca. DM 150.
24. Waterman 452 Î, lever filler, sterling silver, 1926. Value: ca. DM 350.
25. Wahl-Eversharp, lever filler, 1927. Value: ca. DM 60.
26. Waterman 452, lever filler, sterling silver version, 1927. Value: ca. DM 175.
27. Waterman 7, lever filler, 1928. Value: ca. DM 120.
28. Waterman 94, lever filler, 1928. Value: ca. DM 150.
29. Parker *Duofold Junior Lucky Curve*, pressure filler, 1929. Value: ca. DM 50.
30. Parker *Duofold Junior Pencil*, 1929. Value: ca. DM 40.
31. Carter *Superwear*, lever filler, 1930. Value: ca. DM 250.
32. Waterman *Patrician*, lever filler, 1930. Value: ca. DM 500.
33. Conklin *Endura Nozac*, piston filler, 1930. Value: ca. DM 120.
34. Wahl-Eversharp *Doric Gold Seal*, lever filler, 1931. Value: ca. DM 130.
35. Conway Stewart, lever filler, 1933. Value: ca. DM 50.
36. Sheaffer *Lifetime Pen White Dot*, lever filler, 1934. Value: ca. DM 90.

37. Montblanc 6, push filler, 1935. Value: ca. DM 100.
38. Osmia *Baby*, piston filler, 1935. Value: ca. DM 35.
39. Parker *Vacumatic*, membrane filler, 1935. Value: ca. DM 50.
40. Parker pressure pencil, 1935. Value: ca. DM 40.
41. Sheaffer *Lifetime White Dot*, lever filler, 1932. Value: ca. DM 120.

42. Onoto *The Pen*, lever filler, 1937. Value: ca. DM 50.
43. Sheaffer *Lifetime Crescent*, lever filler, 1938. Value: ca. DM 180.
44. Waterman *Hundred Year Pen*, lever filler, 1940. Value: ca. DM 250.
45. Waterman *Hundred Year Pencil*, 1940. Value: ca. DM 140.
46. Wahl-Eversharp *Skyline*, lever filler, 1940. Value: ca. DM 60.
47. Wahl-Eversharp *Skyline*, 1942. Value: ca. DM 75.

German Manufacturer's

1. Pelikan 100, piston filler, blue marbled, 1930. Value: ca. DM 100.
2. Pelikan 101, piston filler, white-gold plated, 1931. Value: ca. DM 250.
3. Pelikan 112, piston filler, 14-karat gold covering, 1932. Value: ca. DM 400.
4. Pelikan 100 *Toledo*, piston filler, sterling silver, 24-karat gold, 1931. Value: ca. DM 400.
5. Pelikan 100 N, piston filler, Kroko pattern, 1938. Value: ca. DM 100.
6. Pelikan *Ibis*, piston filler, 1936. Value: ca. DM 40.
7. Pelikan *Rappen*, piston filler, 1932. Value: ca. DM 70.
8. Pelikan 100, piston filler, 1932. Value: ca. DM 50.

9. Pelikan 400, piston filler, 1950. Value: ca. DM 75.
10. Pelikan 400 N, piston filler, 1955. Value: ca. DM 75.
11. Pelikan 500 N, piston filler, cap and filler grip of rolled gold, 1956. Value: ca. DM 100.

12. Montblanc, lever filler, 1930. Value: ca. DM 30.
13. Montblanc No. 4, safety filler, 1920. Value: ca. DM 80.
14. Montblanc 232, piston filler, cap and barrel of 14 karat gold, 1934. Value: ca. DM 600.
15. Montblanc *Meisterstück 122 S*, push filler, 1936. Value: ca. DM 70.
16. Montblanc *Reflex*, pressure filler, 1936. Value: ca. DM 30.
17. Montblanc 242, piston filler, 1950. Value: ca. DM 65.
18. Montblanc *Meisterstück 139*, piston filler, 1938. Value: ca. DM 250.
19. Montblanc *Meisterstück 142*, piston filler, 1950. Value: ca. DM 120.

The collection of writing instruments is still a very young hobby. The number of collectors has grown sharply in the last ten years. One finds the collector-by-chance, who keeps a handful of pens from his father's or grandfather's times in his desk, likewise the collector-for-fun, who buys what he likes for their color or form at flea markets. The more advanced collector has already specialized. He looks for pens of one type, one brand, or one era. Dedicated collectors of this type have their collectors' clubs. The best-known clubs are:

Pen Fanciers' Club
1169 Overcash Drive
Dunedin, Florida 33528, USA

Writing Equipment Society
c/o Harry Sharf
175 A. Ulverley Green Road
Solihull, West Midlands
892 8AA UK, Great Britain

Belgian Pen Collectors' Association
M. P. van Hoof
Boite Postale 71
B-1000 Bruxelles 23
Belgium

Club des Collectionneurs de Plumes
7 Rue des Epinettes
Paris 75017, France

Internationales Forum
Historische Bürowelt e.V.
Uwe H. Breker
Markusweg 10
D-5000 Köln 50, West Germany

Imaginative new creations of old types of fountain pens are made by Waltraud Bethge at Eppendorfer Landstrasse 122, Hamburg. Michael Pflüger resides at the same address. This talented goldsmith produces writing instruments, jewelry and key rings in his "Workshop for Everything Individual."

Among the world's most outstanding collectors are Hiroshi Nakazono of Japan and the American Mel Strohminger. As a young man, Mr. Strohminger encountered a Parker jack-knife fountain pen by chance and developed an incredible passion for historic writing instruments. Today he is the possessor of more than 10,000 pens.

The German Hans-Georg Schreiver-Abeln of Suhr, near Bremen, has made a name for himself as a collector of ballpoint pens. Since 1983 he has collected ballpoint pens from all over the world, about 6500 of them.

The most famous and oldest shop for old writing instruments is probably "His Nibs," owned by Philip Poole, 182 Drury Lane, London WC 2B 50 L, England. He owns the world's most extensive collection of steel pens, numbering more than 5000 different types.

Whoever would like to keep up on the latest news from the pen collectors' market will find plenty of information in the international collectors' journal *Pen World*, Post Office Box 6666, Kingwood, Texas 77325-6666, USA.

20. Kaweco, safety filler, sterling silver, 1920. Value: ca. DM 150.
21. Kaweco, pressure filler, 1920. Value: ca. DM 50.
22. Kaweco *Sport*, piston filler, 1936. Value: ca. DM 50.
23. Kaweco *Dia 90*, piston filler, 1935. Value: ca. DM 50.
24. Kaweco, piston filler, 1941. Value: ca. DM 30.
25. Kaweco *Sport*, piston filler, 1960. Value: ca. DM 40.
26. Osmia, safety filler, 1923. Value: ca. DM 30.
27. Osmia 66, piston filler, 1933. Value: ca. DM 80.
28. Faber Castell 36, pressure filler, 1938. Value: ca. DM 60.
29. Faber-Castell *Gentleman*, piston filler, 1950. Value: ca. DM 40.
30. Faber-Castell, piston filler, 1950. Value: ca. DM 30.
31. Soennecken *Seneca*, pressure filler, 1928. Value: ca. DM 20.
32. Soennecken 1035, silver-plated safety filler, 1924. Value: ca. DM 120.
33. Soennecken *Rheingold 613*, pressure filler, 1938. Value: ca. DM 40.
34. Soennecken 120, piston filler, 1947. Value: ca. DM 50.
35. Soennecken *Medicus*, pressure filler, 1935. Value: ca. DM 40.
36. Soennecken 105/2, piston filler, 1935. Value: ca. DM 30.
37. Soennecken *222 Extra*, piston filler, 1950. Value: ca. DM 45.
38. Soennecken 111, piston filler, 1952. Value: ca. DM 80.
39. Geha *Regent 710*, piston filler, 1950. Value: ca. DM 60.
40. Geha *Standard 750*, piston filler, 1950. Value: ca. DM 40.
41. Geha *Goldschwinge*, piston filler, 1961. Value: ca. DM 60.
42. Greif *Junior*, piston filler, 1936. Value: ca. DM 40.
43. Lamy 27, piston filler, 1955. Value: ca. DM 40.
44. Brause school pen, piston filler, 1952. Value: ca. DM 15.

A blend of writing instrument museum, sales shop, and repair shop is "Leos/Laden" in Vienna, Universitätsstrasse 6. The proprietor is Leo Grahofer, who earned a bit at the Vienna Flea Market during his student days. He is still found there at times today, and after a "find" presents a display of the pens that will be lovingly restored and offered for sale in his shop. His usual stock amounts to about 200 pens at prices from 600 to 70,000 Austrian Schilling.

As a student too, Reinhard Boettger of Hannover began to specialize in old writing instruments. He is not only capable of restoring almost any brand to writing condition, but he also repairs the old synthetic materials, bakelite, galalite and celluloid, almost invisibly. In addition, he also produces his own imaginative creations from an almost incalculably large supply of original parts for a great variety of brands. Original machines and equipment that he was able to purchase from a Heidelberg fountain pen manufacturer help him.

More than 5000 different steel pens and an assortment of historic fountain pens, kept on hand by Philip Poole in London.

20
21
22
23
24
25

26
27
28
29
30

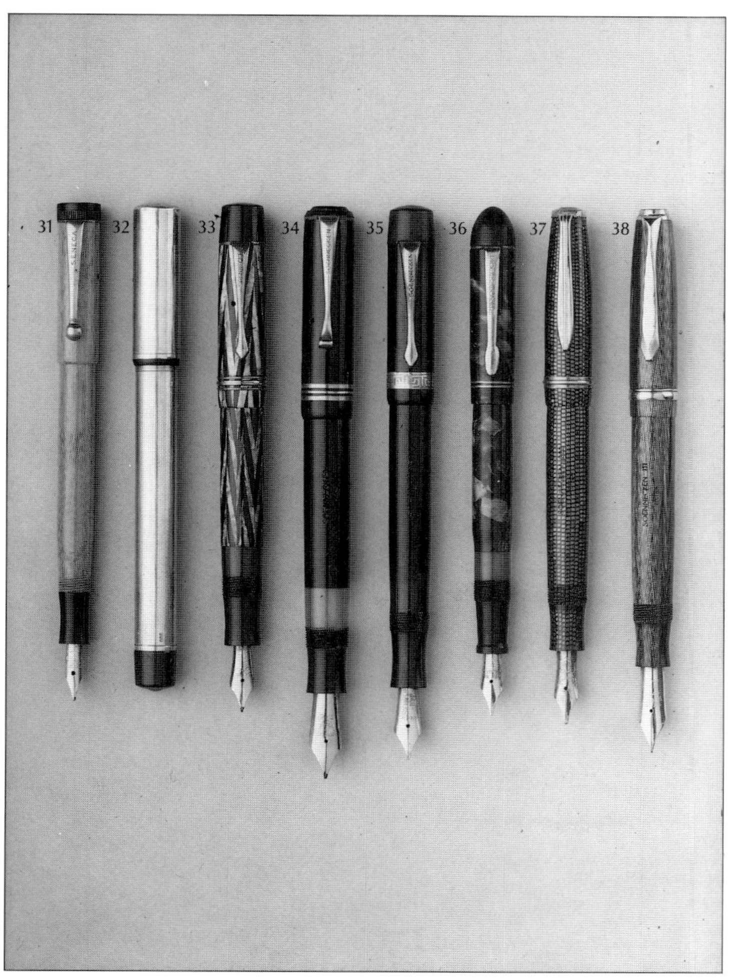
31
32
33
34
35
36
37
38

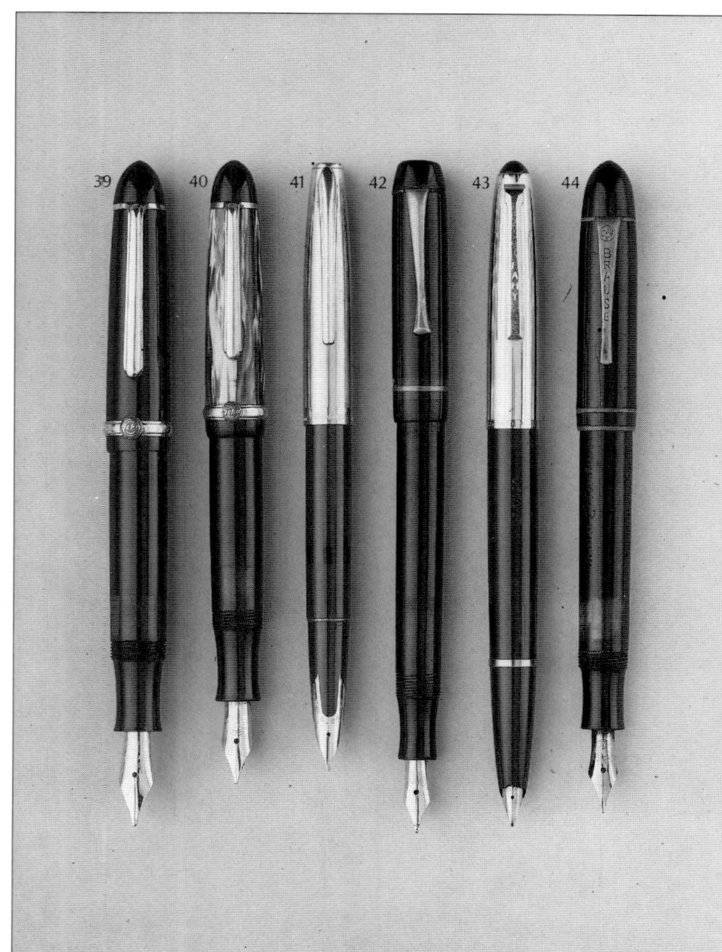
39
40
41
42
43
44

The Fountain Pen—

"The man who wants to get the best results must use the best tools!"
by Johann Wolfgang von Goethe

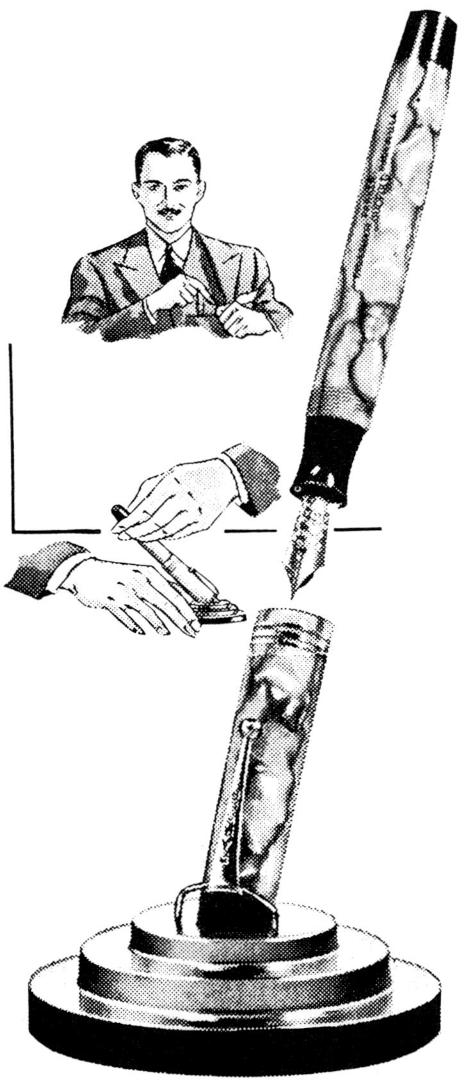

A parking place for fountain pens. One slipped the clip into a mount. The pocket fountain pen turned into a desk pen. Parker, 1932.

A 1918 advertisement.

Just creating a very ordinary fountain pen for Tom, Dick and Harry, that writes properly and lasts a long time, is not an easy job. For many inventive brains that was not enough. They wanted to create something special, types of writing instruments that promised to meet as many different personal needs as possible. From the pen for everybody there soon developed writing instruments for very specific groups of users. In 1932, for example, if the portable fountain pen was not impressive enough, one could get a *Penparker* whose cap would fit into a pen stand and thus turn itself into a stationary desk pen during office hours. Pelikan, Parker, Sheaffer, Waterman and others offered special devices with removable lengtheners, for use on pen stands of impressive onyx.

In 1915, pens that did not need to be supplied with ink were offered by Onoto and Parker for the soldiers in World War I. Camel did the same in the 1930s. An ink tablet was placed in the barrel, and one filled the pen with water when one wanted to write. For more than a year, dry cartridges of this type were supposed to provide their own fresh ink again and again. The firm of Salz in New York cleverly developed pens the same size and shape as rifle bullets: the *Army* and *Navy* pens.

The English Whytwarth *Jumbo* of 1920 was known as a "doctor's pen" because of its size and reliability for use by country doctors in writing prescriptions, as did the 1927 OMAS *Penna del Medico* and the 1903 *Physicians Fountain* by Parker. In the OMAS and Parker models, a small thermometer could be taken out after a screw cap was removed.

Only a fountain pen was not enough? Conklin's *Ensemble* of 1940, Schnell's *Penselpen* of 1929, and other special models were equipped with a mechanical pencil at the far end of the barrel. The *Europa* in 1935, offered two pens for different writing widths were offered in one instrument. This double fountain pen was only exceeded by the OMAS *Colorado* in 1945. It consisted of two longitudinal fountain pen halves that, when separated, could offer two widths and two different ink colors in one instrument.

Everybody has probably encountered the pen with a built-in digital clock. Whoever would have thought that Waterman was already making a fountain pen with a built-in winding clock in the 1920s? This pioneer of fountain pen history also produced such items as fountain pens with built-in lights, especially for reporters, stenographers, mining engineers, and maybe even spies.

Numerous manufacturers of writing instruments seem to have had spies in mind. In 1940 and 1945 Ronson made fountain pens that included cigarette lighters. British aviators could be equipped with fountain pens that included five miniature compasses, dye to color uniforms, aspirin and pervitin tablets, as well as ink for writing . . .

The most expensive secret writer of all time has surely been included in the *Guinness Book of World Records*: the solid gold Montblanc that Roger Moore, as James Bond, used as a lifesaving weapon in the film *Octopussy*.

Whoever wants to be less secretive but more practical can still buy the gold ring ballpoint pen from which a complete rubber stamp with his signature folds out.

Fountain pen and pencil in one item. Schnell's Penselpen was a handy combination for the traveling salesman.

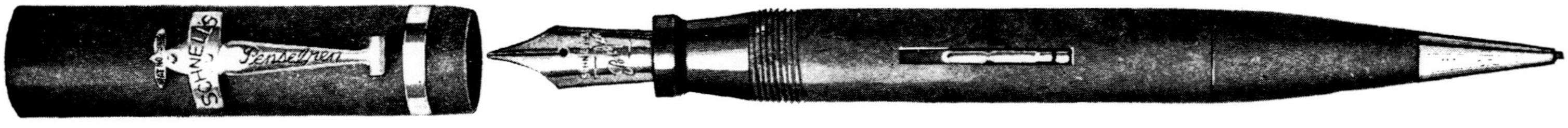

The Materials

The first more or less handmade fountain pens of the nineteenth century were made of wood, metals, natural horn, bone, ivory and mother-of-pearl. L. E. Waterman turned his first *Ideal Pen* out of the hickory wood of a broken wagon-wheel spoke. For series production, though, the natural product, wood, was not very suitable. The required precision for drilling the ink passages, cutting the screw threads and making close-fitting parts could not be attained with this brittle, easily breakable natural material.

Hard Rubber

Around the turn of the century the technology of caoutchouc had already reached a high level. By adding larger portions of sulfur and fillers, the soft raw rubber could be turned by vulcanization into an excellent product for the production of writing instruments: hard rubber, also called ebonite or vulcanite for the sake of elegance.

Pyralene brought glowing color to the world of writing instruments, which had been monotonously black till then.

Kaweco's 1911 assortment for special purposes.

The preferred supplier of caoutchouc was the city of Para in Brazil, from which the raw material came in football-sized balls. The caoutchouc was reduced in size, washed, and dried for several months. Then it was rolled into sheets and vulcanized by adding sulfur. Rolled on thin, smooth steel cylinders, the hardened rubber spent several hours in a steam vat. Then the hollow hard rubber tubes were removed and cut on lathes into the correct sizes. Then the inside and outside threads and decorative wave or line engravings were added.

Waterman ornamented the smooth barrels and caps with jewel-like Youth Style filigrees of silver or gold. Soennecken in Germany was one of the first manufacturers to offer metal sheaths of silver, 14-karat gold, or gold plating.

Because of its special qualities, ebonite remained the preferred material for the fountain pen industry until well into the 1930s. The hard rubber, impervious to the acids in the ink, could be drilled, turned, milled, engraved, and given a high polish with a mixture of oil and rottenstone. The increasing needs of the fountain pen manufacturers and the bitter competition inspired the caoutchouc industry to make further developments. By adding suitable pigments, Waterman was able to offer its ripple models in 1923 in red, olive green, pink, and turquoise.

But under the strong effect of heat and light, hard rubber tends to disintegrate. The smooth black outer surface turns to a dull olive brown, the material cracks and then crumbles easily.

Galalite

The artificial horn called galalite was one of the first modified natural materials to be used in the production of writing instruments. The process, invented in 1897, uses casein obtained from skim milk as its raw material. Under the effects of heat and pressure the mass becomes plastic, and coloring agents give it a marbled effect. It can also be made transparent by adding alcohol. To harden it, galalite has to stay in a formaldehyde bath for several months, which makes production very expensive and time-consuming. These disadvantages were more than made up for by the advantages. The colored material can be polished wonderfully to a high gloss. Its glowing colors, firmness, and elasticity, as well as its ivory-like sound, allowed its use for good imitations of tortoiseshell, coral, turquoise, marble, and amber. Along with its use in the pen industry, galalite was also used to make combs, buttons and piano keys. A further important advantage: galalite is easy to solder. Damaged pens could be softened in hot water or acetone and easily repaired. Its disadvantage: Over the years it lost its treasured surface gloss, absorbed moisture, and broke easily.

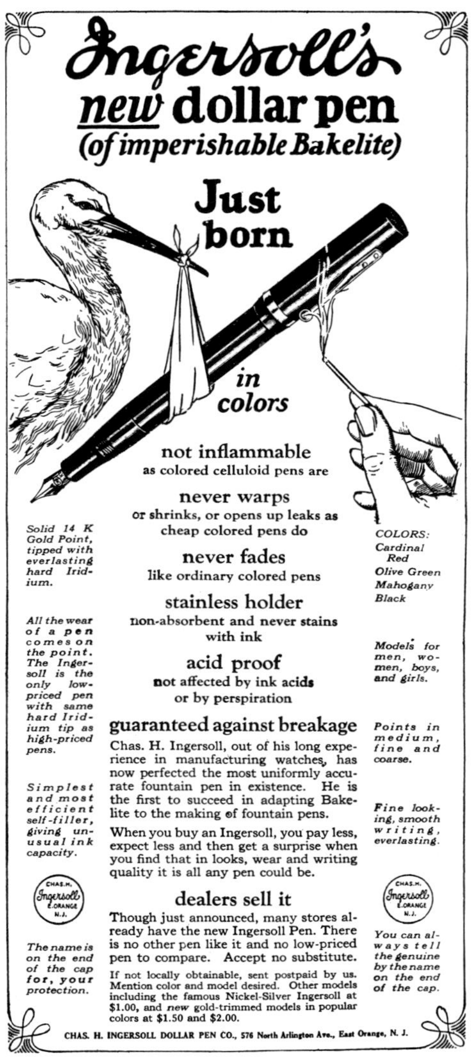

Fountain pens of imperishable Bakelite, 1928.

Bakelite

Bakelite is the first artificial resin, invented by L. H. Bakeland. It was made by condensing phenoles and formaldehyde, and has been used in pen production since the beginning of the twentieth century. Because of its qualities of being unmeltable, insoluble, and insensitive to heat, pressure, hitting, or bumping, and of being as clear as glass, Parker used Bakelite for the first time in 1916 to make the legendary pocket knife fountain pen, the *Jack-Knife*, with its transparent barrel. Since Bakelite requires a laborious pressing process in matrices, it was not to prevail as a standard material for pen production.

Celluloid

The oldest of all modified natural materials is celluloid. It was invented by Hyatt in 1896 as a substitute for ivory in the manufacture of billiard balls. Celluloid consists of a solid solution of cellulose nitrate and camphor as a softener, and in a heated condition it takes any form, which it retains when cooled. Celluloid can be stamped, blown, welded, cut, scaled, and sawn. Like practically no other artificial material at the beginning of the 1920s, celluloid was suitable for production in any desired color. Thus plates of different colors were cut into pieces measuring about two millimeters, mixed and pressed together in a press. Objects with tortoiseshell patterns were made by pressing transparent yellow and dark brown celluloid sheets three millimeters thick. Similarly to galalite, celluloid can be brought to a high gloss with polishing devices or rotating cloth discs.

In 1924 the American chemical firm of Dupont developed a synthetic material similar to celluloid called pyralene, using cellulose nitrate plus a softening agent. The public was excited by the possibility of being able to buy fountain pens and mechanical pencils in glowing colors or marbled, striped, flamed, or speckled patterns. Sheaffer was the first to make use of this opportunity, calling the material "Radite." Parker followed with "Permanite," which it used for the legendary unbreakable *Duofold*.

Celluloid or pyroxylene remained the synthetic material most frequently used in the writing instrument industry until the 1940s, and was still used in Germany in the 1960s. While the public loved celluloid because of its brilliant colors and unbreakable nature, the manufacturers feared celluloid. Its ingredient of guncotton was also the raw material for the manufacture of explosives. Since its kindling temperature was only 170 degrees, higher temperatures could and did cause explosive fires. That was probably the reason why Montblanc, for example, singled out "fireproof pens" for praise in their 1924 price list, as not being ". . . made of cheap celluloid and inflammable."

Pelikan and Parker are probably the only manufacturers who still offer writing instruments like those of the 1930s and 1950s. Of course the easily inflammable celluloid has long since been replaced by safe cellulose acetate.

Lucite

Pyroxylene caused a problem for Parker at the end of the 1930s. It was decided to offer a quick-drying ink for the Parker 51. The first experiments showed that the alkaline components of the ink dissolved the pyroxylene of the ink reservoir. Looking for a new synthetic material that was impervious to ink and unbreakable, Parker came upon a material that had proved itself in the American armaments industry. The Air Force used a synthetic material called "Lucite" for aircraft noses and turrets.

Thermoplastics

Approximately at the beginning of the 1950s, the age of plastics began. Thermoplastics such as polystyrene, ABS, PVC, and plexiglas could be shaped when heated and produced quickly and economically in large quantities in special multipurpose injection tools. Modern plastics have special qualities, depending on the type, such as unbreakability, flexibility, and resistance to aging and light. In terms of their surface gloss and glowing colors, they are not at all inferior to their predecessors galalite, bakelite, and celluloid. The opportunity offered by the injection process for thermoplastic materials is that twelve or more parts can be produced simultaneously in one injection process. This is simultaneously an advantage and a disadvantage. The high investment for the construction of the injection machinery reduces the variety of forms and sizes to just a few models. Unlike celluloid and galalite, thermoplastics can generally be produced in only one color.

Metals

Gold and silver have been treasured since antiquity not only for their beauty. These precious metals also possess material advantages such as being malleable, workable, and durable. Gold does not corrode. In a pure state, gold and silver cannot be worked. The metals are too soft for that. Therefore they are mixed with copper, zinc, nickel and other metals. The proportion of gold is measured in karats.

24 karats means 1000 parts of gold, 18 karats = 750, 14 karats = 585, 12 karats = 500, 10 karats = 416, 8 karats = 333 parts. The three most important shades of gold are: Yellow gold = 3/4 gold and 1/4 silver with the same proportions of copper; red gold = 3/4 gold, 1/4 alloy with little silver and much copper; white gold = 3/4 gold, 1/4 alloy of nickel or palladium.

Pen caps and barrels of solid 14-karat gold are rare and valuable articles. Their genuineness is attested to by stamped lettering on the barrel, cap, nib and sometimes also on the clip. For reasons of expense, the writing-instrument industry usually turns to covering processes. Thin sheets of brass or copper alloys, which are covered by even thinner sheets of 10-to 18-karat gold, are known as rolled gold or gold doublé. The karat count of rolled gold is stated as 14 or 18 karats, gf or GFL. The karat count in itself tells little; the thickness of the gold covering is much more important. It measures from 3 to 12 microns. One kilogram of rolled gold sheet often contains no more than 20 grams of gold.

To produce the barrel and cap, round pieces of sheet metal are drawn into cylindrical shells in a succession of work processes, using stretching tools and hydraulic presses, or made out of tubes by hammering machines. Gold-plated means that the metal parts are given a coating of gold in an electrogalvanic bath. As a measure of quality, one differentiates between "gold-plated" (= at least 2.54 microns), "electroplated" (= at least 0.18 micron), and "gold flash" (= 0.05 to 2-micron thickness).

Silver, almost as desirable as the more valuable gold, must likewise be alloyed, usually with copper, to be usable for jewelry or writing instruments. The usual silver alloy is sterling silver. It consists of 925 parts of silver and 75 parts of copper. The name "sterling" comes from the old English silver penny, in the middle of which a star was stamped. The people referred to it as a "starling," little star, and this later became "sterling." During the course of time, the weight of a pound of silver pennies (sterling) was used to determine the weight of the English pound. Today the concept has been separated completely from the English currency and is used exclusively for the silver alloy. Along with the usual stamped number 925 for sterling silver, the numbers 825 and 800 are also found, indicating smaller proportions of silver.

Just as with rolled gold, there are also platings and coverings of silver. For "silver sleeve," an external tube of 925 sterling silver with a thickness of about 0.1 millimeter is slipped over a sheet-metal tube. Only solid silver and "silver sleeve" may be stamped as 925 silver. German silver ("new silver" in German) is an alloy of copper, silver sulfide, nickel and zinc.

Unlike gold, silver discolors. Particularly high dampness and gases in the air discolor silver. For this reason, silver articles are covered with a protective coat of lacquer after production.

Stainless Steel

Stainless steel is one of the modern materials used in the writing instrument industry. It is extremely scratch-resistant. Certain alloys are extraordinarily corrosion-resistant. Its high elasticity and unbreakability make stainless steel especially suitable for nibs and clips, and for caps and barrels as well. Stainless steel can be finished matt, sandblasted, or polished to a high gloss.

Engraved Metals

All precious metals and high-priced alloys, as well as copper, brass, and stainless steel, can be engraved. Metal is removed with the help of the rotating point of an engraving machine, by hand, or with very modern laser beams. Very thin layers of alloys such as rolled gold are given a material-compacting "Guilloche" that prevents the gold layer from being perforated and the underlying brass or copper material from becoming visible. This machine process was developed by the

Frenchman Guillot at the end of the 17th Century. Working from a pattern, a needle feels the prepared model and transmits its form via a rod to the guilloche needle that copies the pattern on the material being worked.

Diamond Cut

Diamond cut is an impressive name for a simple procedure. It means nothing more than the material-removing working of a plastic or metal tube with a diamond as the cutting edge. The hardness of the diamond produces a faultlessly smooth, highly polished surface that does not need to be polished afterward. The diamond cut also makes artistic facetings possible.

China Lacquer

China lacquering is one of the most artistic and expensive processes of decorating the surface. The lacquering technique has been known to the Chinese for thousands of years; its secret was guarded and carefully passed on from one generation to the next. The raw material is the natural lacquer "urushiol," taken from the resin of the sumac tree. It is worked by hand into as many as ten layers, one on top of the other. Every layer is individually dried and then carefully polished. Thus a high-gloss, externally scratchproof surface is built up, layer on layer.

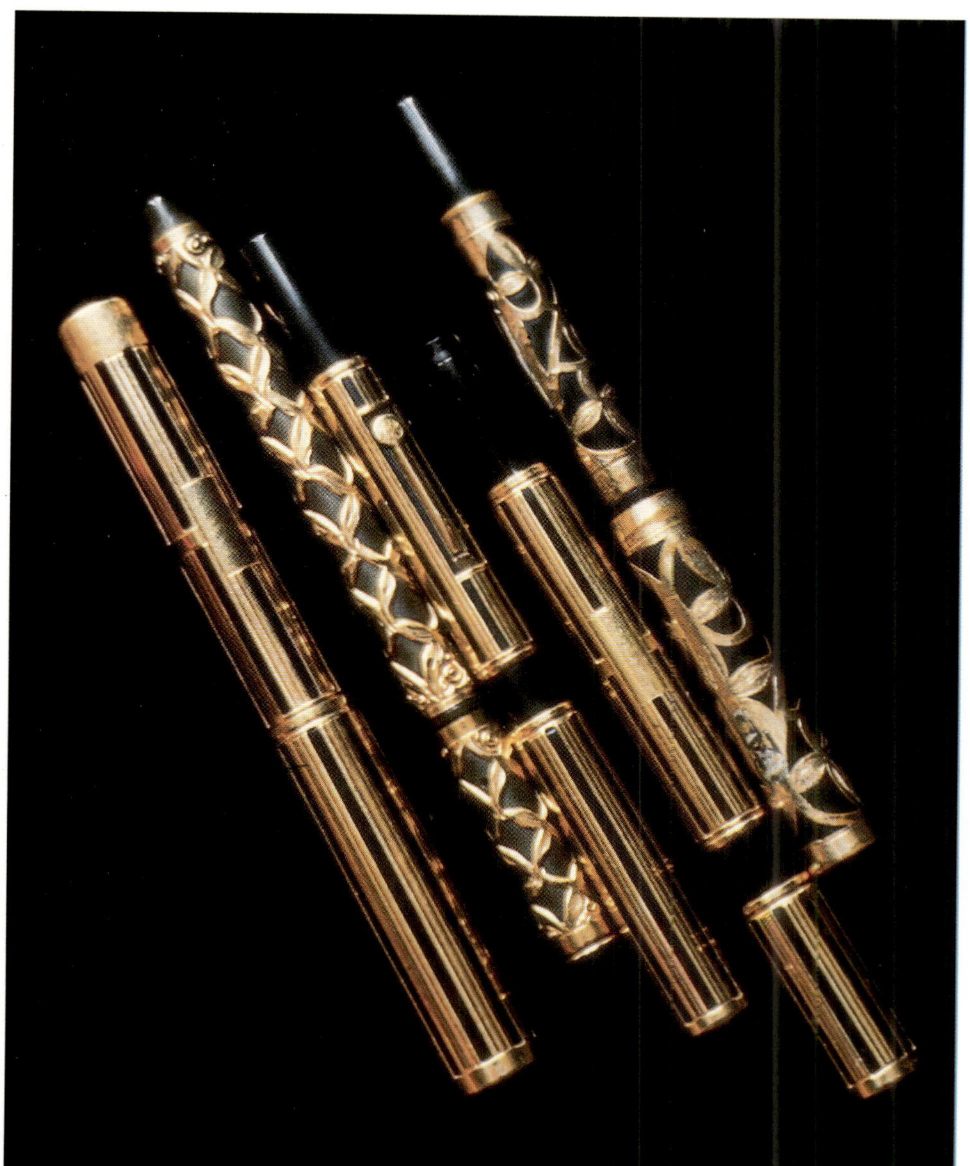

Gold or silver filigree decorated these Waterman models.

Thomas de la Rue (Onoto, Pelican)

The origins of this old English manufacturer of writing instruments go far back. The printer and publisher Thomas de la Rue had already made a name for himself as the producer of excellently printed and colorful playing cards when the introduction of the "penny post" in England in 1840 made the demand for envelopes increase dramatically. Until this time it was customary to fold letters for mailing and seal them with wax. In 1851 de la Rue introduced the first machine to make envelopes automatically. Thomas de la Rue soon became the best-known and busiest manufacturer of the new envelopes, as well as printing bank notes and postage stamps. Completing the growing assortment of paper goods with a writing instrument was an obvious next step.

Starting with a patent by T. R. Hearson, Thomas de la Rue developed the "Anti-Stylograph" in 1881, a crude instrument with a covering of hard rubber. An ink valve let ink reach the paper whenever the point was pressed firmly onto the paper, not an ideal system for a reliable fountain pen. An improvement followed in 1884. The *Swift* had a turning bar in the ink reservoir with which the pen could be closed again. In 1895 the *Pelican* was introduced in Harrods' catalog as the first non-blotting pen that could be carried in any position and supplied itself. In addition, the ink flow could be regulated very accurately by turning the barrel on the front section.

The *Pelican* was a great success for many years, until the popularity of the rising self-filler forced the development of a more reliable pen. A gigantic advertising campaign introduced "Onoto—The Pen" in 1906. The name, an artificial word, had the advantage of being pronounceable in any language. The spectacular Onoto possessed a removable piston mechanism with which one "cannot only fill instantly, but can also dampen the ears with water, spray flowers with plant-protective materials, and be used to drive off pests on the street by spraying them with ink." What more could one ask of a pen?

The Onoto was followed in 1910 by the *Hermit*, a safety filler, and the *Pentago* with a five-sided cap that could not roll off the table. Both pens had the ink passage under the nib for the first time.

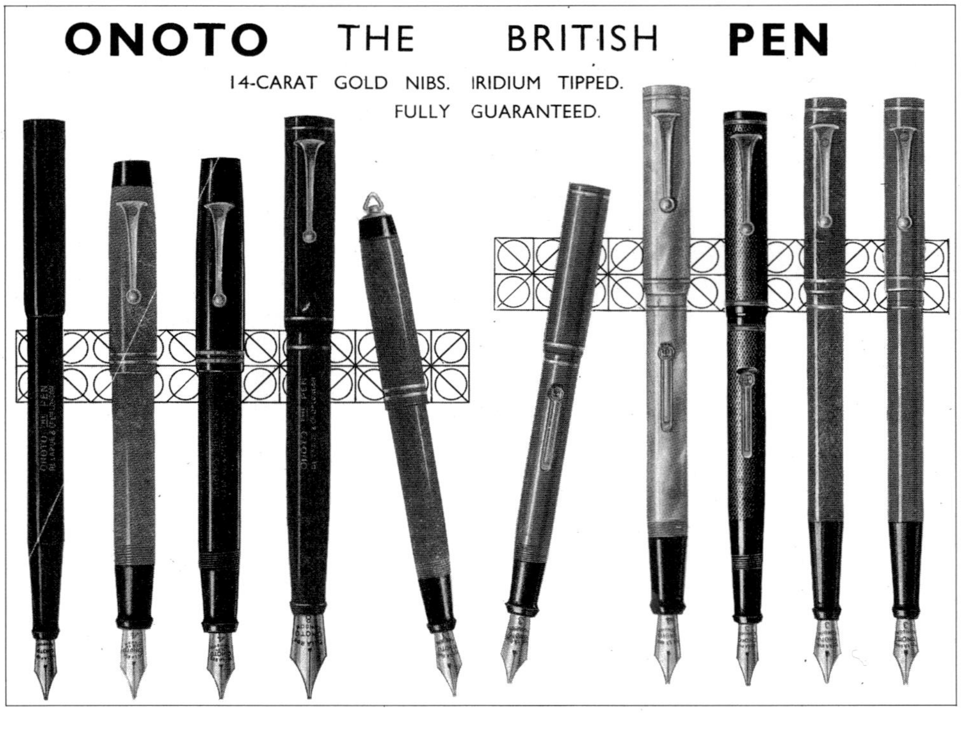

A page from a 1930 brochure.

Just before and during World War I, the firm of Thomas de la Rue could proudly call itself England's greatest manufacturers of fountain pens. In 1921 the situation changed. The company fell into financial difficulties and moved its factory from London to Scotland. Although they were convinced of the advantages of the piston filling system, the firm had to bow to the pressure of competitors who had long produced a wide range of lever fillers. So as not to get the reputation of being an old-fashioned producer of pre-war products, the company developed a new lever filler and produced the Onoto in the new, originally American celluloid colors of green, blue, pearl, and black. Some of the new products could already be offered with a transparent barrel for a better view of the ink supply.

At the end of the 1930s, Thomas de la Rue probably had the largest line of colored, semi-transparent fountain pens: the standard model, *Magna, Ladies' Model, Minor Model, Self-Filler* and the "Ink-Pencil" stylograph. This assortment remained almost unchanged until the end of the 1940s, extended only by the ballpoint pen. At the beginning of the 1950s, the British public's interest in British-made fountain pens decreased. The firm tried to answer the success of the *Parker 51* with the K Series and regain the lead. But the new series resembled the products of its Parker competitors so closely that any real success was impossible.

On February 27, 1958 the firm of Thomas de la Rue & Company Ltd. halted the production and sale of Onoto writing instruments in Great Britain. For a time, the Norman Baker & Longhurst Pty. Ltd. firm produced Onoto writing instruments under license in Australia, small quantities of which also reached the British Isles. But time had finally run out for the onetime giant among manufacturers of writing instruments.

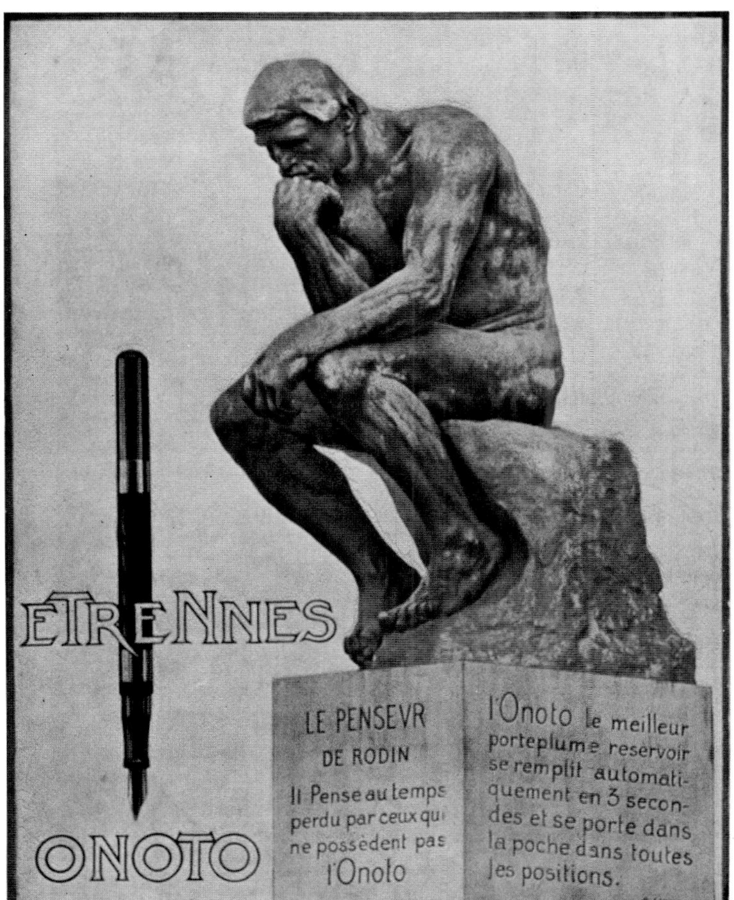

Whoever would have thought it could be filled in three seconds. A 1908 advertisement.

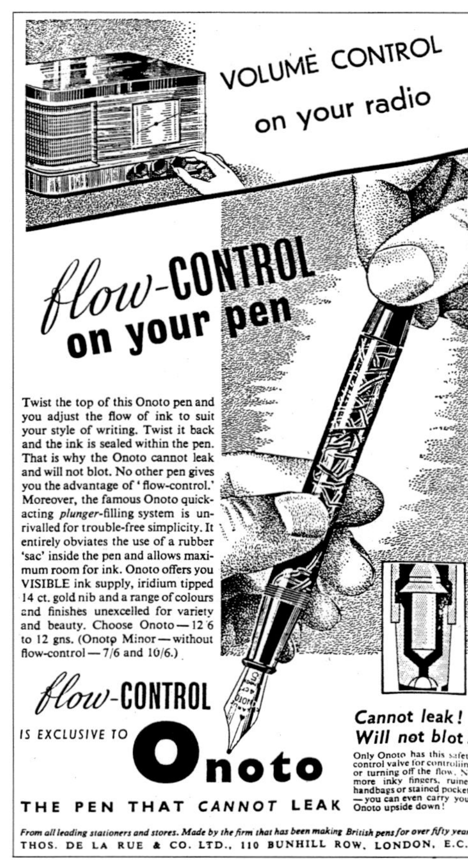

Individual ink-flow regulation by means of a control valve, 1934.

Industrial Reports:

From Raw Metal to Gold Pen.

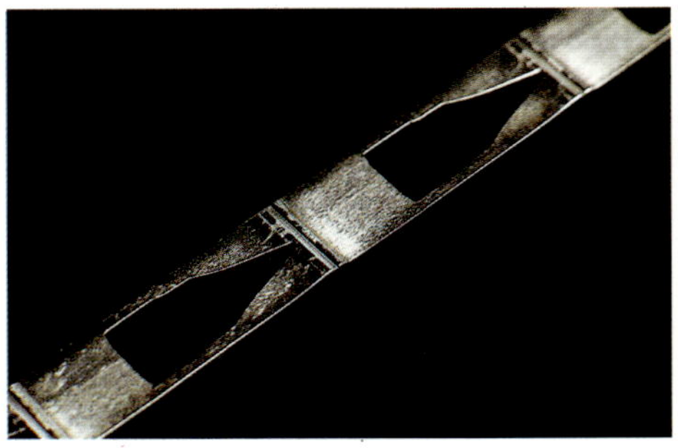

Stamping the points out of a rolled metal strip.

Stamped-out pieces.

Stamping tools.

The osmium-iridium writing ball is welded on.

Smoothing the writing ball below on the rotary grindstone.

Smoothing the sides.

Paper-thin impregnated copper blades for slitting the nibs.

Fast-moving discs cut the slit in the nib.

Smoothing and polishing the surface with water and copper granules in rotary containers.

The surface of the nib is brought to a high polish.

Mounting the ink passage.

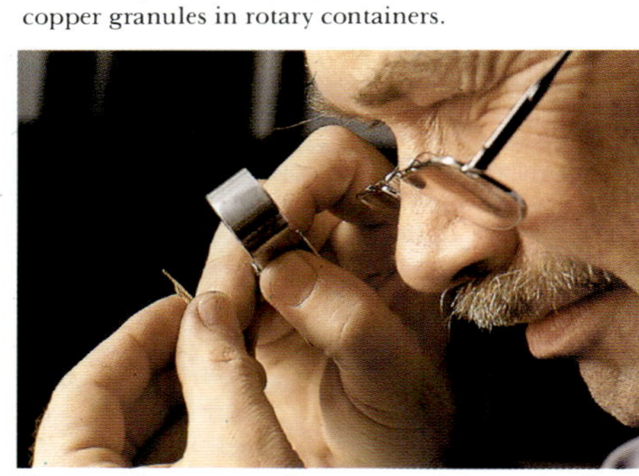

Every work process undergoes a careful examination.

Every pen is broken in "dry" by hand . . .

. . . and every piece is tested mechanically and electronically.

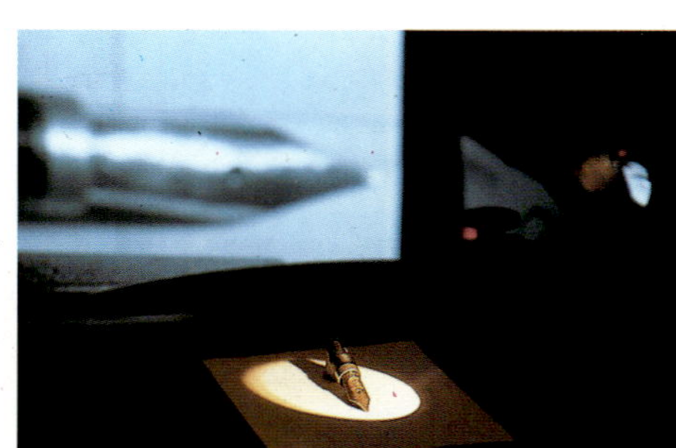

The process of manufacturing a gold nib.

The adjustable Eversharp pen offered nine different nib sizes.

Osmium-iridium—ten times more valuable than gold, to make sure the nib does not change, even after years of use

The Pen Gives Handwriting Character

The nib is the vital part of any fountain pen; it determines the quality of the handwriting. Quality and writing comfort depend on the material and workmanship of the nib. A good nib glides smoothly over the paper and transmits every change in pressure, from the finest line to the widest. Unlike the rigid ballpoint pen, pencil, or felt-tip marker, the fountain pen nib gives handwriting its personality, expressiveness and character.

Writing nibs are made of gold (18, 14, and formerly 12 and 16.67 karats) or stainless steel. Gold nibs for luxury pens often also have a surface decor of rhodium, a precious metal of the platinum group. Stainless steel nibs are also gilded, according to their price range, in thicknesses from .1 to .2 microns.

The production of good fountain pen nibs demands much experience, precision, and skill. Even today, individual handwork is required in most of the work processes.

The basic material is a strip of gold or stainless steel that is cut into rectangular plates. Each plate in pushed through two rotating steel drums that roll the piece of metal thicker at the front and thinner at the end. This establishes the eventual firmness and elasticity of the nib. In the next step, the nib plates are stamped out according to their later point width and finally engraved with the brand name, width, and karat markings and decorative lines. At the same time, a hole is often stamped in the nib as well, to supply the ink passages better with air and guarantee a steady flow of ink.

In another work process, the bending of the nib plate into a trough form takes place.

The welding of the writing ball of expensive, wear-free iridium, osmium, ruthenium or tungsten alloys is a particularly important process. Without this special point, gold and steel nibs would be worn down by writing after a short time and lose their predetermined stroke width. Iridium points can last a lifetime when treated with care. Welding the ball demands the utmost precision and is decisive in the further working and quality of the nib.

After welding, the writing surface of the ball is ground smooth and adjusted to the precise writing width. The splitting of the pen, unlike that of the ancient goose quill, is not longer done with a penknife but with a fast-rotating thin slitting blade.

Now the inner edges of the nib shank and head must be rounded by grinding and polishing, so as the guarantee optimal ink flow. The quality of the nib is dependent above all else on the care with which this work is done.

The almost finished nib is highly polished for hours, sometimes for days, in a rotating drum containing polishing agents, thus taking on its "final smoothing." After being mounted on the ink-passage section of the pen, it can be inscribed. High-priced gold nibs are inscribed by hand. This specialist job demands great sensitivity and manual dexterity, both of which can be developed only through long practice.

Pen Widths

Every person writes differently and thus requires a pen cut to suit his type of writing. The most frequently used nib widths are:
EF, F, M = sharp (extra fine, fine, medium)
B, BB = wide (broad, very broad)
OF, OM, OB, OBB = oblique/left-angled (fine, medium, broad, very broad)
St = steno nib
A = beginner's nib (school pens)
L = left-handed nib

Stenographic nibs have to reproduce the finest lines reliably and react to the lightest pressure. They must also allow writing speeds of 450 to 480 syllables per minute.

There have been many attempts to improve the comfort of fountain pens. At first the manufacturers tried to meet the individual buyers' needs with as wide a range of nibs as possible. For example, Montblanc offered an assortment of 108 different types and sizes in 1936. The smallest nib was only 1.5 cm long, the largest 4 cm.

To allow the use of different nib sizes in one pen, Eversharp produced its "adjustable nib" in 1934. A special mechanism allowed the user to move a control button in the pen to nine different positions and thus attain nine different nib adjustments—for everything from an elegant letter to carbon copying.

Almost all the European and American manufacturers followed the trend toward prestige nib types until the 1940s. The attempt to go against the trend seemed doomed to failure from the start. Yet in 1941 Parker had the courage to try. The construction of the epoch-making *Parker 51* produced a small nib, enclosed in plastic as much as possible, so as to prevent the premature evaporation of the new quick-drying "51 Ink." Naturally such a "mini-nib," of which only the very point was visible, hardly possessed enough elasticity for European taste. Yet the small nib dominated international fountain pen design in the 1950s and 1960s, though the quick-drying ink that had led to its development no longer played a role.

Taste changed in the 1970s. The small nib grew from one model generation to the next. Scarcely anything was still reminiscent of the classic free-standing or covered nib. The writing tips of the new styling lines fit into the design concepts and constituted a formal continuation of the front section of the pen. Noteworthy examples of this are the Pelikan M 30, the Montblanc wing and intarsia pens, the Lamy 2000, etc. Then Parker, and later Pilot, went one step farther. In the Parker TI (1970) and the Pilot MR (1978), the nib and front section consisted of a single piece. Of course the public could not really warm up to the titanium-alloy Parker TI or the stainless steel Pilot. The TI, whose stroke width could be adjusted to fine, medium, or wide with the help of two adjusting knobs under the nib, was given up after only one year of production. Quite a rarity!

Speaking of rarities, in the mid-70s, Pilot introduced the first push-button fountain pen with a retractable point. It combined the advantages of the nib with the handiness of the ballpoint pen. Pressing on the end of the barrel made the nib emerge at the other end for writing. Another push moved it back inside the barrel. The nib of this cartridge filler was protected from drying by a cleverly designed, automatically opening and closing flap mechanism.

Illustrations at upper left, page 137
The most commonly used nib sizes. The enlarged pictures show the specially ground nibs.

Size nib sizes for different handwriting—a 1930 Parker advertisement.

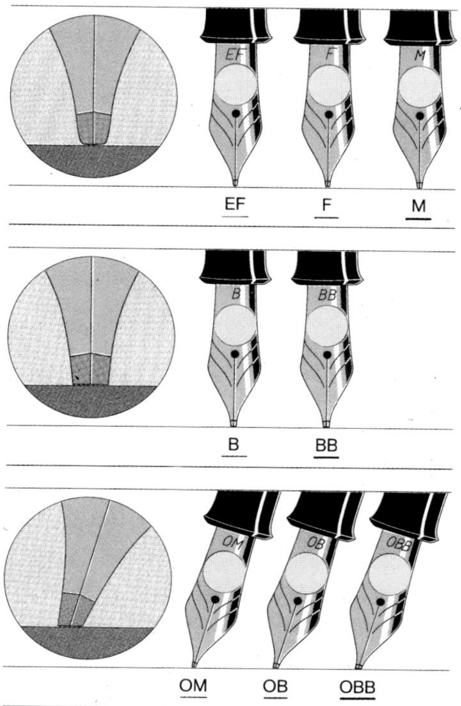

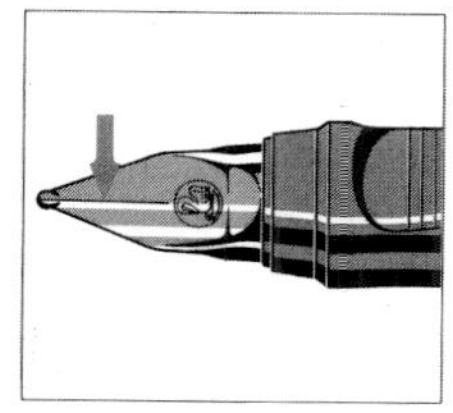

Too-thin ink flow.

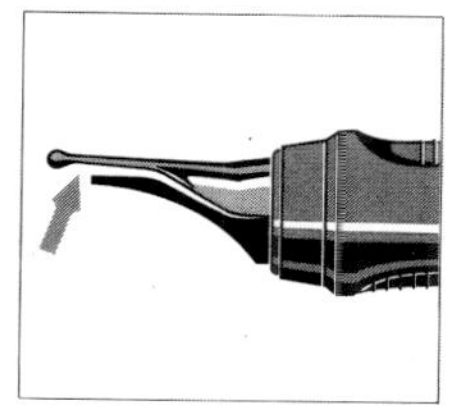

The ink flow stops with too little pressure on the nib.

The nib gives off no ink.

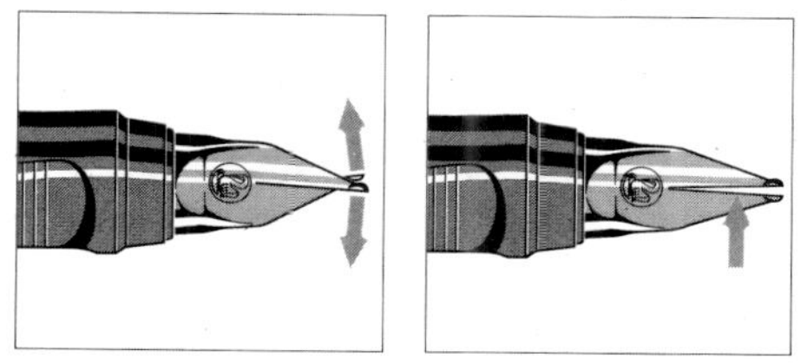

Nib Care

The idea that one should not lend a fountain pen is widespread. The nib could become bent or changed otherwise by another type of writing position. Have no fear! The high flexibility of a gold or steel nib always withstands considerable pressure. The fine writing ball of extremely hard metals related to platinum is scarcely worn by years of use. So a nib can experience no damage through use by a strange hand, unless it is damaged by being hit, bent, or dropped. Bent nibs should always be repaired or replaced by the mamufacturer or a fountain pen shop. Minor repairs can be made oneself with the greatest care and caution, and without guarantee.

Too-thin ink flow

Cause: The nib slit, when held to the light, no longer shows a hairline opening.

Remedy: A sharp, very fine knife blade should be inserted into the slit carefully from below.

Ink flow stops with light pressure on the paper

Cause: The ink passage is separated from the nib. A gap is visible.

Remedy: Press the rounded back of the nib gently on a surface. In most cases, though, the nib must be removed and reshaped.

Nib supplies no ink

Cause: The nib slit is open too wide.

Remedy: Press the nib together from both sides until the tips cross slightly over each other.

Drops of water on the nib

Cause: Condensation through cooling of moist air in the closed cap. A problem that appears particularly in pens that close absolutely airtight and thus do not dry out.

Remedy: Clean out the inside of the cap with a tissue, clean the front section and nib.

The Ink Passage: Heart of the Fountain Pen

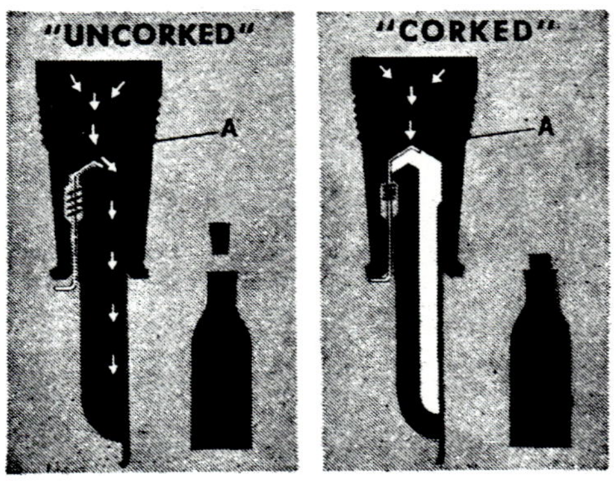

"Corkable" ink passage by Eversharp, 1937.

The fact that ink flows out of a fountain pen is attributable to, among other things, a barely visible, remarkably complex part, the ink feed. The first really useful ink passage was "invented" by L. E. Waterman in 1884. It allowed the ink to flow evenly and undisturbed from the ink reservoir to the nib without stopping or spattering. The secret of a good ink passage is its utilization of the natural capillary action of the ink in very fine channels. Some of the channels draw ink on as soon as the pen touches the paper and ink is given off. . Other channels let fine bubbles of air move in the opposite direction, into the ink reservoir. A well-built regulating passage lets in only as much air into the reservoir as equals the volume of ink consumed.

When the fountain pen was still in its infancy, there were differences of opinion among the manufacturers as to whether the ink feed should end on the nib, below it, or above and below it to guarantee optimal ink supply. The dispute was settled by esthetic considerations. The passages under the nib no longer concealed the sparkling, valuable gold nib. One of the most industrious and inventive developers was G. S. Parker. The "Lucky Curve" ink passage produced in 1894, as well as the "Spear-Head," ranked for decades among the most intelligent and successful examples of American engineering.

Until the introduction of thermoplastic materials produced by injection molding, the ink passage consisted of hard rubber. In dozens of work processes, ink channels and equalization channels had to be cut and smoothed by hand in special operations, sometimes to tolerances of one hundredth of a millimeter. Every form and size of nib requires its own special ink passage to produce optimal performance. Along with ebonite, metal components for the ink passage were also tried. Types such as the "Swallow Pen" and "Swan Pen" had fine silver threads in their ink passages to move the ink out and let air into the reservoir.

Original variations on the theme of the ink passage can be noted in several models: Eversharp produced safety ink regulators in 1937. The four-section piece opened the ink flow when the cap was unscrewed. When the cap was closed, the parts of the ink passage were pushed together and prevented an unwanted outflowing of ink as well as drops of water that might build up overnight on the nib or in the cap through condensation.

The aforementioned Snorkel Pen by Sheaffer had a periscope-like extending pipe in the ink passage, on the principle of the submarine periscope, through which ink was sucked in to fill the reservoir without having to dip the nib and front section into the ink bottle. Geha offers the classic assistance of an ink regulator with a reserve supply. A part of the ink volume can be used only by switching on the reserve system.

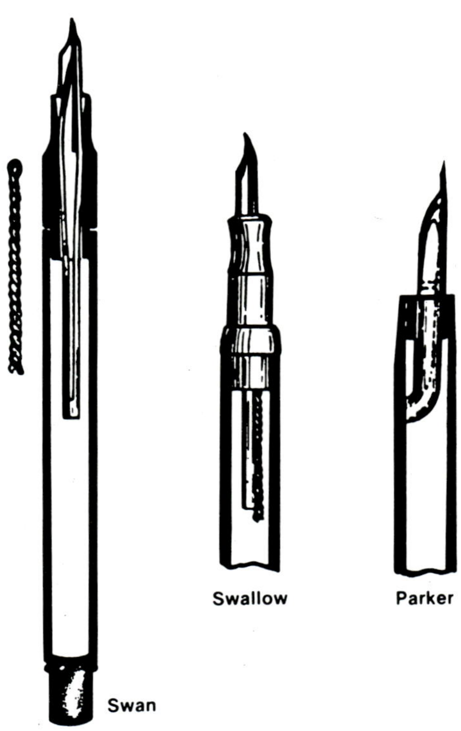

A turned silver wire handled ink flow in some early models.

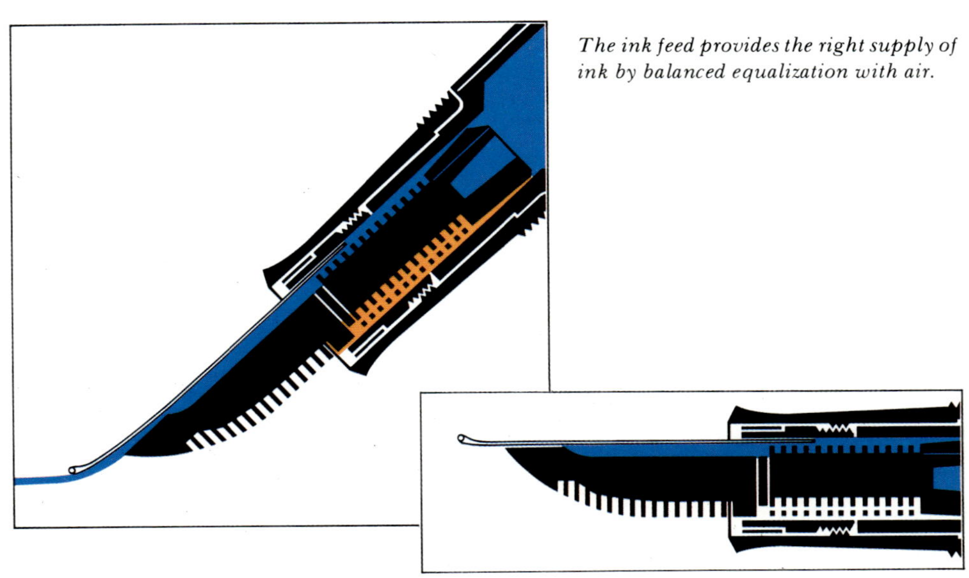

The ink feed provides the right supply of ink by balanced equalization with air.

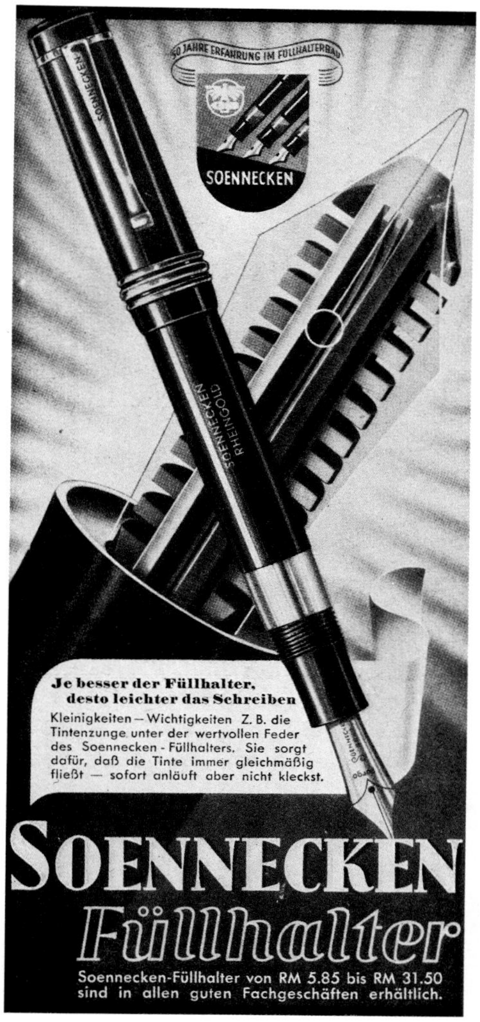

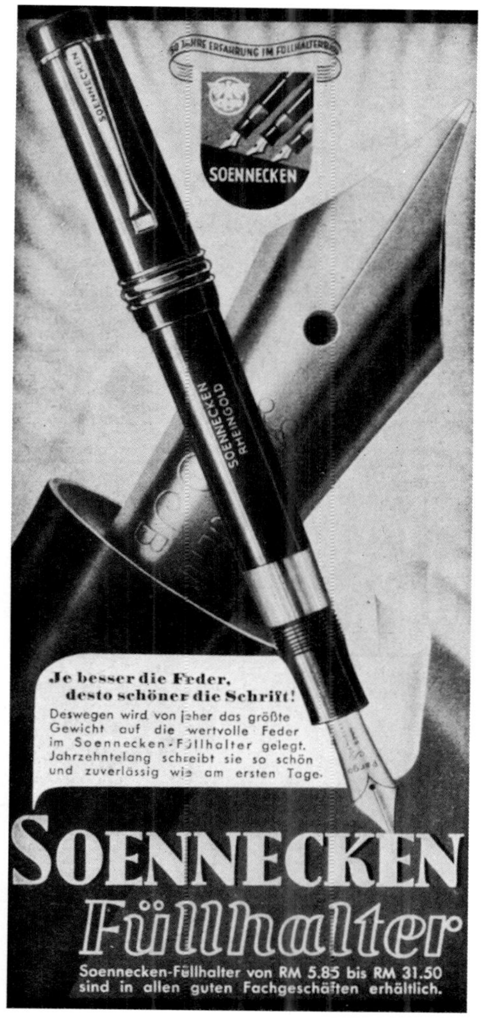

Soennecken particularly stressed the "ink tongue" (ink feed) and nib in this 1936 advertisement.

The Filling System

The search for the perfect filling system has accompanied the history of the fountain pen from the beginning. Every new filling system tried to outdo the previous ones in simplicity and reliability.

The Insertion Pen

The most primitive and simplest filling method required an auxiliary device, a pipette or medicine dropper, such as we still use today for eye or nose drops. With its help one sucked ink from the bottle and let it drip into the unscrewed barrel of the pen. This process required concentration and precision, for it could always happen that the ink reservoir overflowed or the ink flowed out of the dropper prematurely. And naturally the glass eyedropper had to be taken along, on trips, for example. Despite the unhandiness of eyedropper filling, the insertion filler was made until well into the 1920s, especially for export to tropic lands in which the rubber bladder of the self-filler could deteriorate quickly.

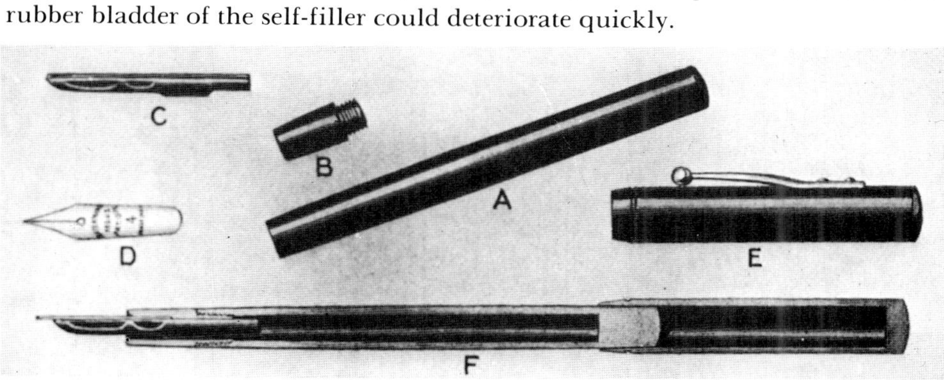

The insertion fountain pen: A = barrel, B = front section, C = ink feed, D = nib, E = cap with clip, F = cutaway view.

The Safety Pen

Another inherent disadvantage of the insertion pen is that it could leak out even when closed. The manufacturers thus recommended that it always be carried in the pocket with the nib upward. For ladies that was an impossibility. They needed an absolutely leakproof fountain pen to carry in their handbags. The "safety fountain pen," produced first by Caw, then by Faber-Castell, Kaweco, Montblanc, Soennecken and Waterman fulfilled this requirement ideally. In its resting position, the nib was inside the ink reservoir. To fill it, one held the pen vertically and unscrewed the top end of the cap. Then the ink reservoir was filled to the brim with an

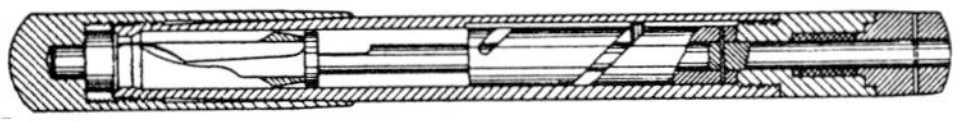

The older type of safety filler with a retracting nib.

Ink tablets of 1916.

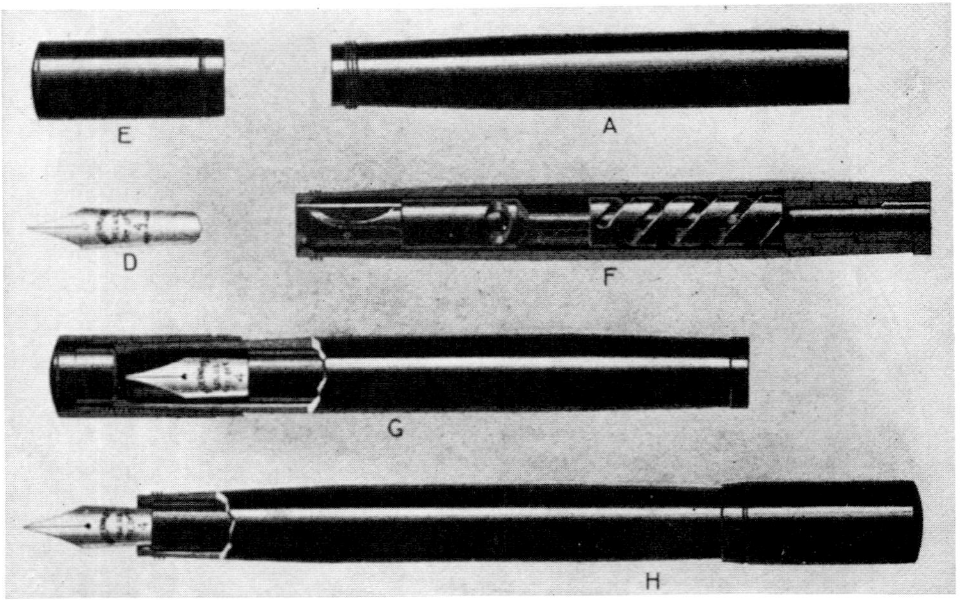

The safety fountain pen: A = barrel, B = nib, C = cap, F = cutaway view of the shaft with snail screw, G = closed pen, H = open pen.

eyedropper. Then one turned the filling handle at the end of the barrel and made the nib extend, which simultaneously closed the ink reservoir and made the pen ready to write. To close the pen, one held it vertical again and turned it to draw the nib back into the reservoir. Now the cap was screwed on firmly. The safety pen prevented leakage absolutely reliably and could be carried in any position. When the pen was closed, the nib was constantly surrounded by ink, and components that would prevent writing could not solidify on it. To avoid damage, for example, if one forgot to retract the nib before screwing on the cap, some types of safety pens such as the Montblanc had a silver bar in the cab.

The Tablet Pen

Several remarkable pens could be used without filling them with ink at all. Filling the reservoir with water was enough to make the Parker Trenchpen (1917), the Aurora Etopia (1934), the 1936 Camel, and models by Onoto and Swan ready to write. All these pens held an ink tablet that dissolved in water and produced a deep black ink.

The Eyedropper Pen

This filling system was surprisingly simple. It functioned, so to speak, like a built-in eyedropper. The barrel end consisted of a long, soft rubber sack. When one pressed this part with two fingers, the air was pushed out of the pen. By pressing it several times, the pen could be filled quickly with ink. Typical representatives of the eyedropper filling system were the Pelikan Rappen of 1932, several American Eversharps and English Wyverns.

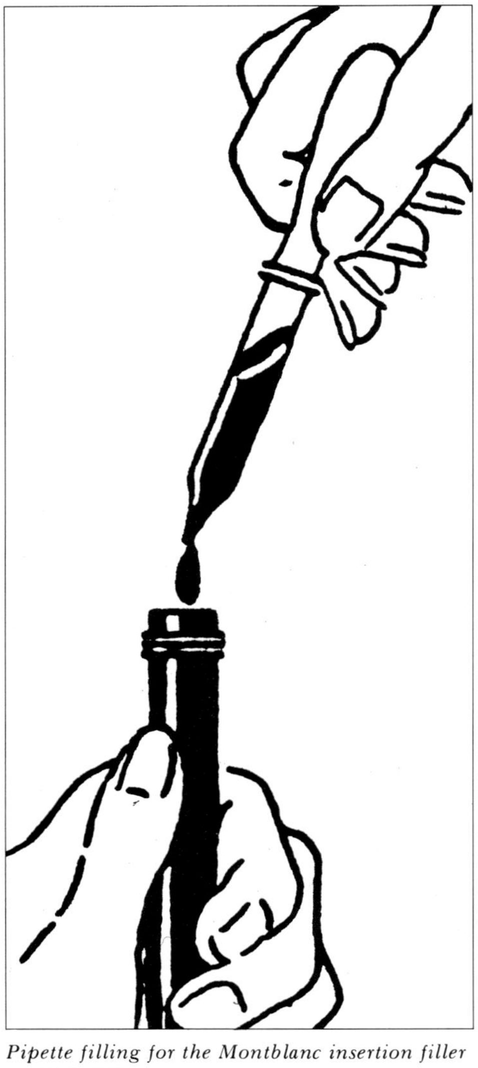

Pipette filling for the Montblanc insertion filler and tube filler.

The Tube Pen

As early as the mid-19th Century inventors were working on the idea of using a flexible rubber tube as an ink reservoir. Kollisch (1844) and Mosley (1859) suggested compressing the tube with a push-button so as to remove the air completely. When released, the tube would fill completely via the ink passage. Unfortunately, the sensitive rubber tube was in danger of tearing or weakening. The Automatic Pen described in 1878 was an attempt to avoid this disadvantage. The metal barrel was to have an opening through which one could press the tube with the thumb to empty it, then release if to fill it.

The Conklin "Crescent" fountain pen offered a further improvement. A metal part shaped like a half-moon projected from the barrel and was connected with a metal rod inside the barrel. A horizontal ring of hard rubber prevented unintentional activation of this lever. The "Crescent" made the self-filler a success. The

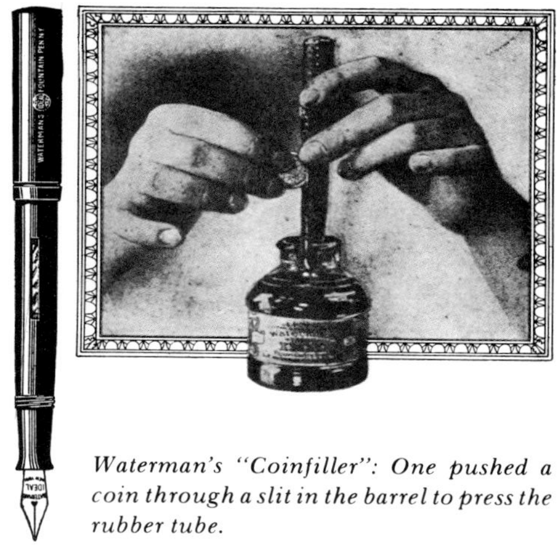

Waterman's "Coinfiller": One pushed a coin through a slit in the barrel to press the rubber tube.

A 1920 advertisement.

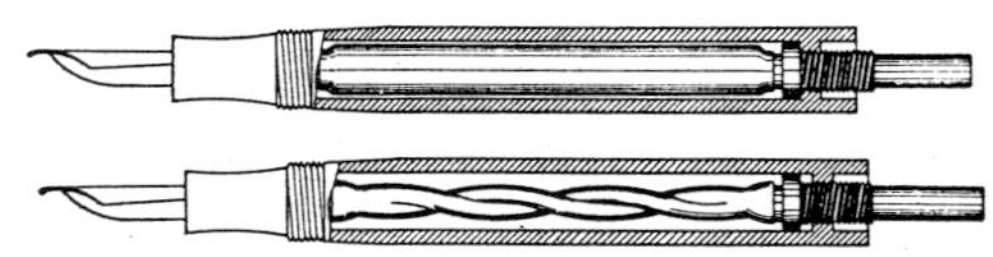

Filling the pen by turning the tube.

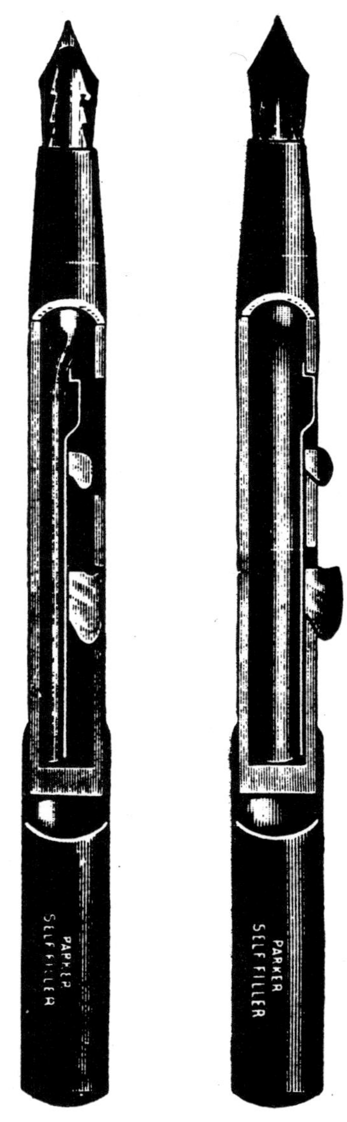

Parker's first tube fillers.

Dr. Faber's first "Twist" fountain pen.

The disadvantage of the safety filler—it was not ready to write immediately—was used in Parker's duofold advertising.

typical appearance with the shining gold ring on the shaft made the Conklin so popular that even famous writers and frequent users such as Mark Twain ranked among its most loyal fans.

The Conklin's success inspired Waterman to adopt a similar filling system in 1912. He avoided the metal crescent piece on the barrel, though; instead he asked the user to press a coin into a corresponding slit in the shaft to pump ink into the tube.

In 1914 the "tube filler" followed. Instead of pressing a knob or using a coin, one pushed an auxiliary tube on the barrel up and down to press the rubber tube together.

Parker approached the solution from another angle. In 1905 he patented a self-filler that had two finlike raised sections on the barrel. When one pressed these "buttons" forward and down with one's fingernail simultaneously, the rubber tube was compressed. To fill the pen, one pushed the buttons in the opposite direction. An audible clicking indicated that the filling procedure had stopped.

Aside from the Conklin pen, none of the early tube fillers achieved a lasting success.

The Lever Pen

The tube filler achieved its final breakthrough in 1907 with W. A. Sheaffer's idea of replacing the buttons, tubes, half-moon and thump pressure with a simple lever. The lever was connected by an axle with a metal rod in the barrel. Externally the lever fit into the smooth surface of the barrel without any disturbing transition. To fill the pen, one pulled it out with a fingernail, the metal rod pressed against the rubber sack and pressed the air out. Filling took place automatically by pushing the lever back into the barrel. One could not imagine filling a self-filler more neatly and conveniently at the beginning of the Twentieth Century. Sheaffer's success was correspondingly great, and the firm soon became one of America's leading producers of fountain pens. The simple construction of the lever filler also convinced the Europeans.

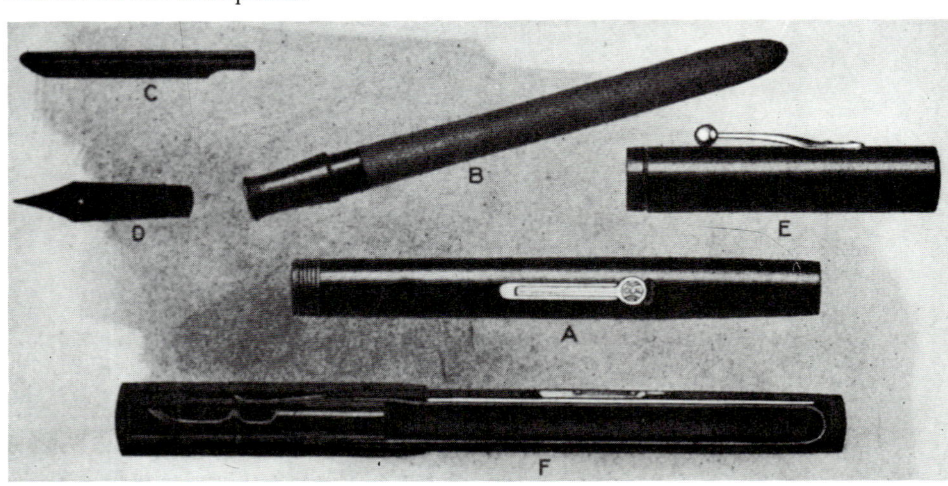

The lever fountain pen: A = barrel with lever in resting position, B = grip section with rubber tube, C = ink feed, D = nib, E = cap, F = cutaway view.

At the beginning of the 1920s, Soennecken followed with a wide range of lever fillers. Waterman, Swan, Thomas de la Rue, Wahl-Eversharp, Conway-Stewart, Montblanc, Pilot, Platinum, Sailor and even Conklin adopted this filling system over the years. Until well into the 1950s, models were produced in France, Italy and Great Britain. Germany was among the few "tubeless" lands in which the lever filler could not find acceptance. Variations on the lever filler were the twist type offered by Dr. Faber in 1906, Schaaf in 1903, Waterman in 1903 and Swan in 1933. A hook or button on the end of the barrel moved a curved panel inside the barrel that pressed or released the tube after a brief turning.

The Pressure Pen

An excellent alternative to the lever filler in general use was the pressure filler. Introduced by Parker in 1916, it soon found use in pens made by Montblanc, Soennecken, Kaweco and many other firms. In the pressure filler, a well in the barrel for a lever could be dispensed with. To fill it, one unscrewed a cap from the end of the barrel and pressed a thin metal rod that pushed a metal rod against the rubber tube to push the air out. About ten seconds after the filling rod was released, the rubber tube was filled with ink. Soennecken eliminated the easily lost cap in its "one-hand" pressure filler. A white knob on the end of the barrel had to be pushed four or five times after a safety ring was opened, in order to fill the pen.

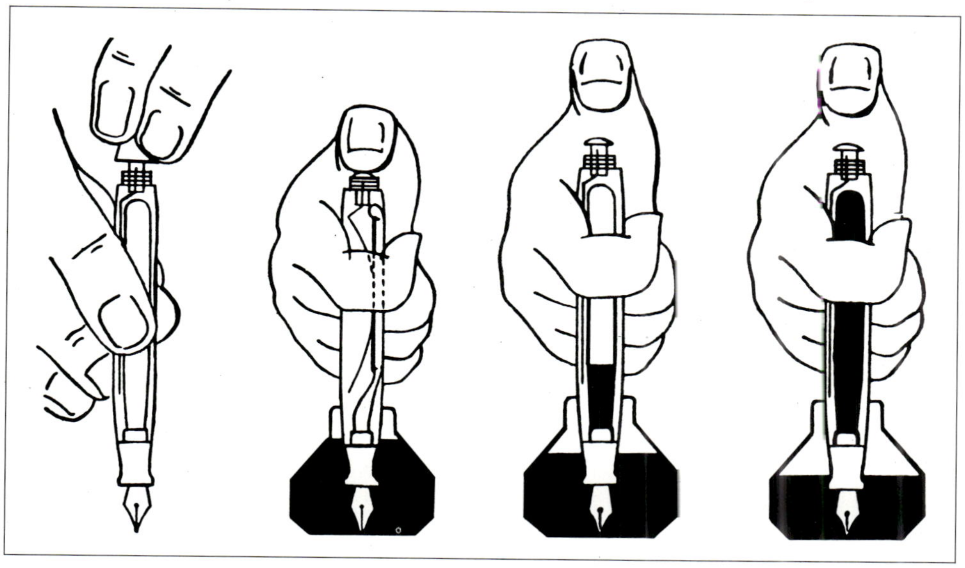

How the Parker pressure filler works: Unscrew the end cap of the barrel, press the button, release it and count to ten—the pen is filled.

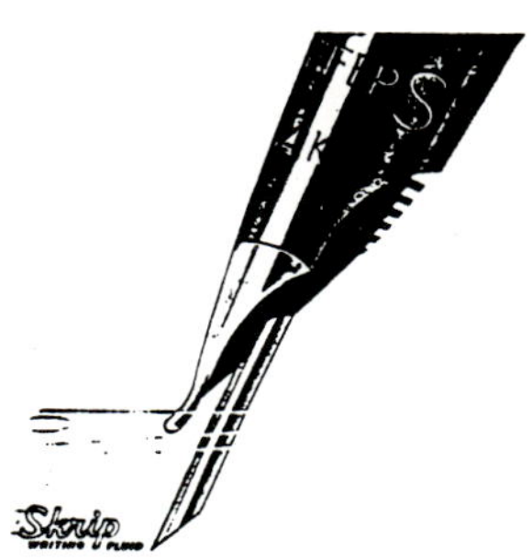

The "Touch-Down" filling system by Sheaffer shortened the filling process.

The long wait until the pen was completely filled was shortened by the push fuller. In the case of Montblanc, one push on the screw end of the barrel was enough to fill the pen. Sheaffer's "Touch-Down" of 1949 functioned as follows: When a telescopic inner barrel tube was drawn out, the rubber tube was compressed by the resulting high air pressure. When it was pushed back in, a groove on the end of the tube let the air pressure decrease—and the rubber tube expanded and sucked up ink.

The Piston Pen

Several manufacturers tried to solve the problem of quick, neat ink filling in a stiff, mechanical way. The Post pen of 1900 offered a simple piston principle. After the barrel was unscrewed, the piston was pulled back by a lengthening rod to fill the pen. A simple procedure, it had the advantage of being able to eliminate a rubber tube that was easily cut and prone to deterioration by the acid in the ink. There was, though, the danger that in time the seal and piston would loosen and the ink would leak out.

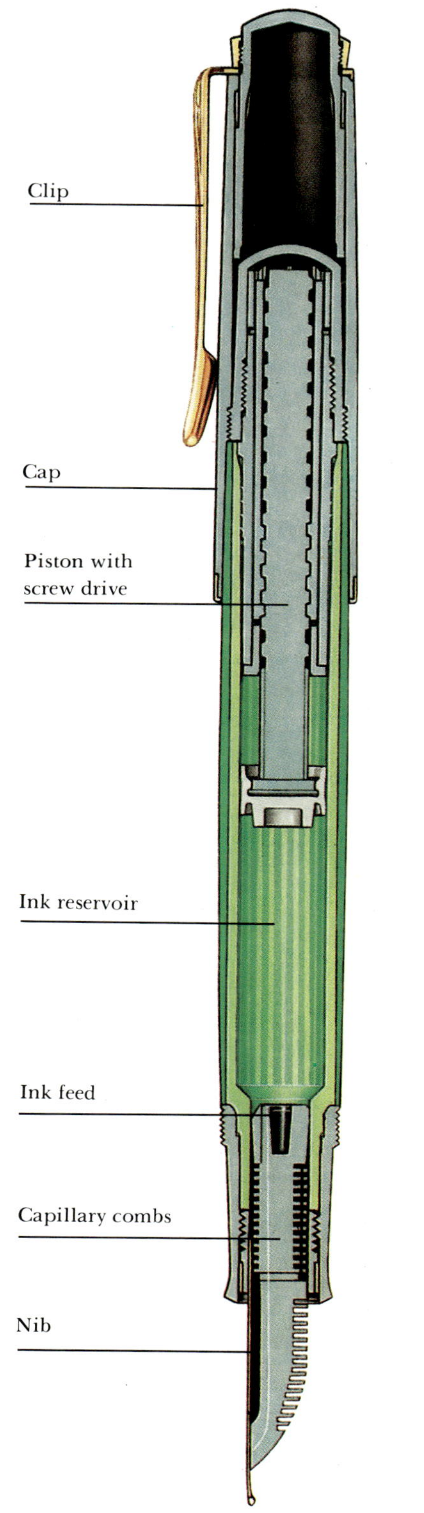

Piston filler with differential screw drive.

Onoto refined the primitive piston push-pull system by replacing the piston rod with a turning knob at the end of the barrel. Sheaffer and Eversharp used a cleverly created vacuum behind the piston to suck up ink at one stroke. The Italian "Universal" even had a pumping system in which the nib had to be pushed against the bottom of the ink bottle to take up ink, a process that required particularly stiff nibs.

The further development of the piston filler remained the domain of the Europeans. In 1929 Pelikan succeeded in producing an absolutely reliable piston pen. A differential screw gear, activated by a filling knob on the end of the barrel, made the piston of fine cork slide up and down in the transparent ink reservoir. At the same time, the reservoir was cleaned. The filling knob's gear shifted into neutral to prevent ink from being forced out while loosening the filling knob. A

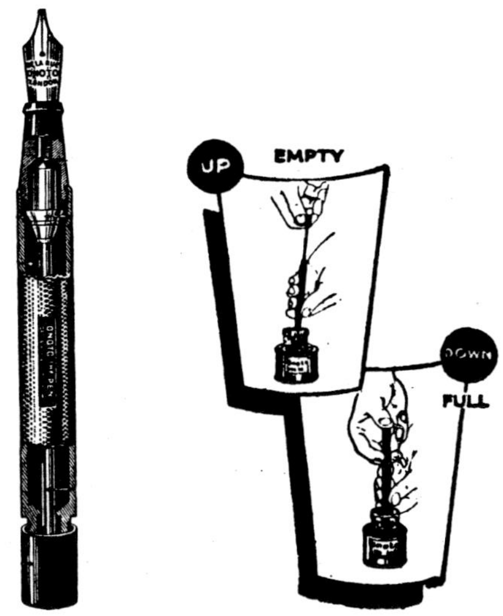

The push-pull piston mechanism of the Onoto fountain pen.

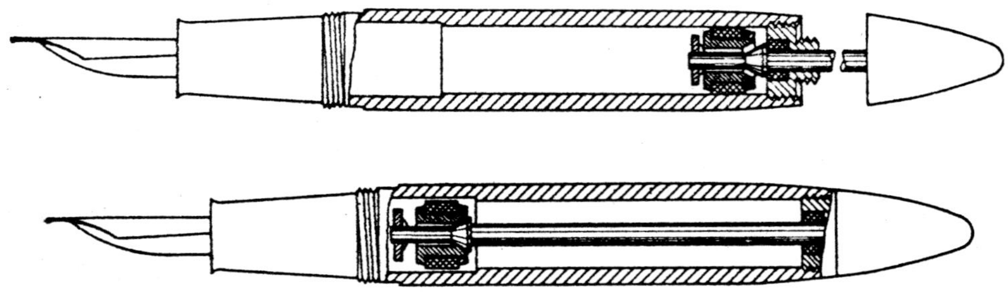

Sheaffer and Eversharp refined the comparatively simple piston mechanism.

double gasket of hard rubber protected against leakage in case the piston itself should become loose.

The piston filling system had great advantages: fast suction caused by the low pressure created when the piston was drawn back, a large supply of ink, long life compared to the sensitive rubber tube, plus a viewing window that let one check the ink level at any time. Until well into the 1960s, the piston filler remained the favorite and best-selling type of fountain pen in Germany. Along with Pelikan and Montblanc, Soennecken, Faber-Castell, Osmia, Kaweco, Staedtler, Geha, Brause and many others included piston fillers in their assortments. In America, only Conklin tried to make use of the European piston-filling system with the "Nozac" in 1931.

The Membrane System

With the *Vacumatic,* Parker astounded the pen world in 1933 with a completely new and remarkably effective filling system. Developed on a scientific basis by Professor Dahlberg at the University of Wisconsin, it united the functioning of the piston mechanism with that of the tube filler. A push-button at the end of the barrel controlled a piston inside the barrel. This piston was covered by a flexible rubber sack. When the push-button was pressed several times, a vacuum was formed inside the barrel. The ink could enter the reservoir through a small tube set in the ink passage.

Despite the relatively laborious and time-consuming pumping, the *Vacumatic* had particular advantages. More than twice the ink supply for longer writing and constant knowledge of the ink level were the selling points with which Parker advertised the new filling system. The only German manufacturer to use the *Vacumatic* filling system was Osmia.

In 1936 Waterman adopted the vacuum principle for the *Ink Vue*: Instead of a push button, the tried and true lever mechanism was used to create a vacuum in a shortened tube in the ink reservoir.

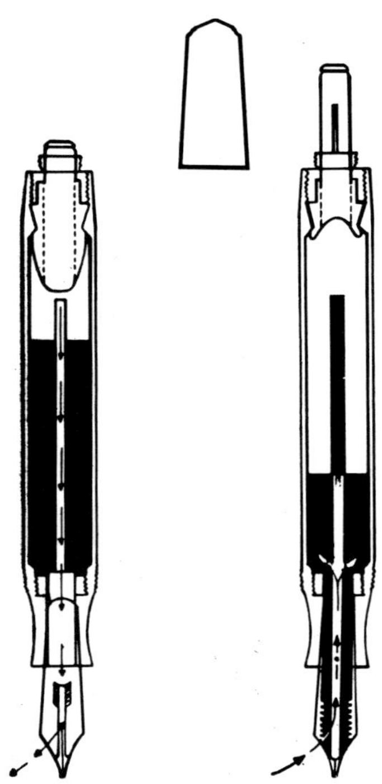

Cross-section of the Parker "Vacumatic." Left: When the button is pushed, air leaves via the ink feed under the nib. A vacuum results in the barrel. Right: After the filling button is released, ink is sucked up through a tube set into the ink feed.

The Nylon Bag

In 1948 Parker returned to the simple "automatic" filling system of the early years of fountain pens. When the revolutionary Parker 51 appeared along with the

Filling without a mechanism. Capillary attraction of the ink in the Parker 61.

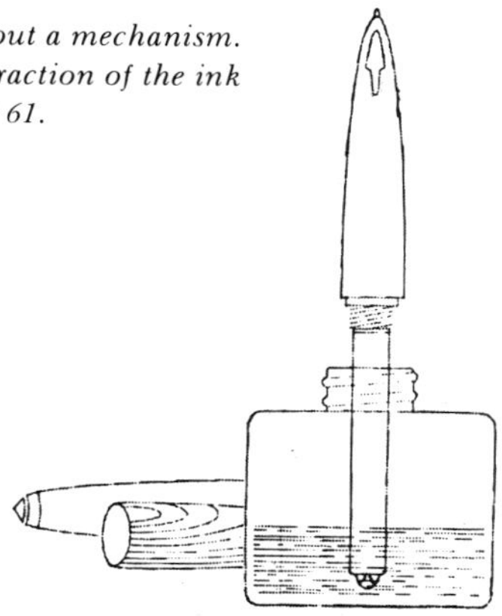

Still in use today: the thumb-pressure aerometric filling system.

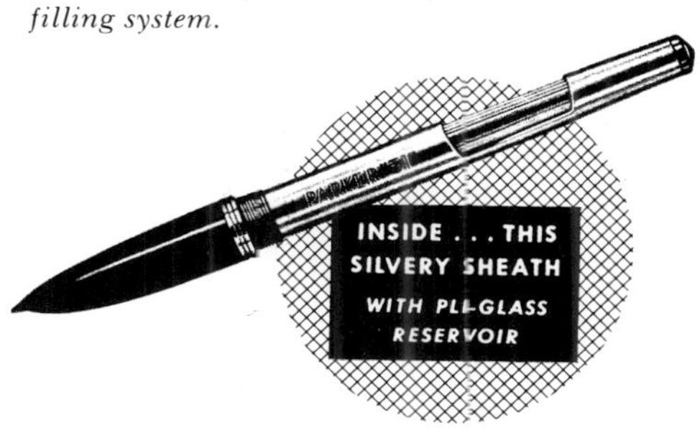

refined *Vacumatic*, the customers had to make do with a more primitive but nevertheless effective system—the flexible ink sack. The further development of synthetic materials allowed the use, instead of a rubber sack, of an almost indestructible bag of Pli-Glas (PVC). This synthetic bag is protected by a metal covering. A side opening about the size of a thumb allows the ink sack to be pressed down, as was done with the fingers around the turn of the century.

The Capillary System

An attempt was made to eliminate sensitive moving parts completely in models like the Parker 61 or the early Jif-Waterman school pen in the 1950s and 1960s. Their ink reservoirs were filled with glass balls of varying sizes or absorbent material similar to cigarette filters. If one dipped the absorbent material or the nib in ink for a time, capillary action drew the ink in. The advantages: The pen filled itself automatically, so to speak, within ten seconds. The pen was long-lasting and inexpensive to make. The disadvantages: The pen dried out quickly and experienced decreasing ink-flow power.

The Ink Cartridge

It is regarded today as the cleanest, handiest, and safest filling system. It was first produced industrially by Jif-Waterman in 1935, in partnership with the American Waterman firm. The small glass cartridge plugged with a stopper, though, did not catch on. Only with the help of modern injection technology was Jif-Waterman

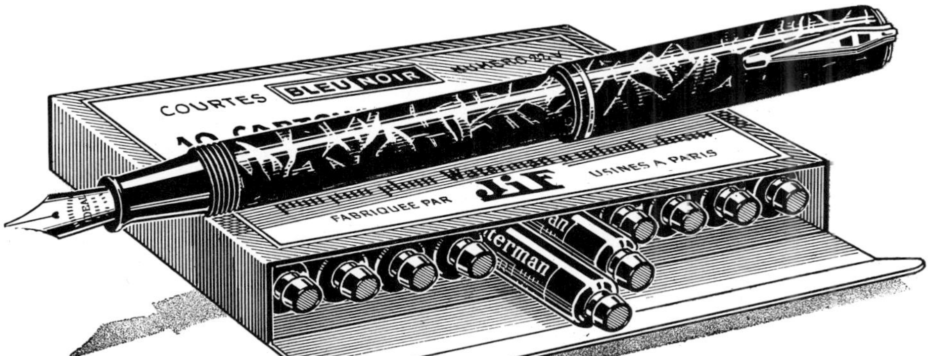

The ink cartridge was invented by Jif-Waterman in France. After the patent expired, this practical system was adopted by nearly all manufacturers.

able to succeed with the ink cartridge in 1953. After the patent ran out, Sheaffer followed in 1955, as did Parker, Diplomat, Pelikan and others in 1960. The ink bottle was retired from service for the time being.

Ink cartridges can be carried anywhere. The European standard reserve cartridge can be put into a fountain pen in seconds-an advantage that students and most adult users would not want to be without. Along with the standard cartridge (used by Pelikan, Geha, Montblanc, Senator, and Reform), large-size cartridges (Lamy, Parker, Waterman, and Sheaffer) are also used.

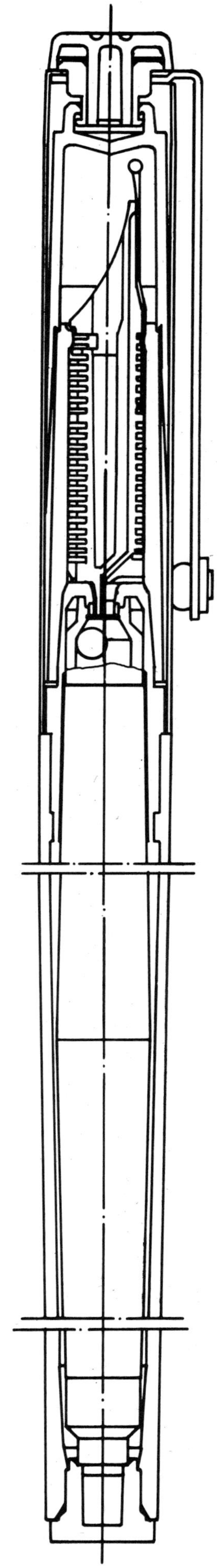

Cross-section of a modern cartridge filler with a reserve cartridge in the barrel.

Mabie, Todd & Co. (Swan)

The English writing-instrument manufacturer, Mabie, Todd & Co., began as an American firm. In the firm, which was founded in New York in 1843, two men made pencil boxes and pens in a small workshop. In 1873 the Mabie brothers, H. H. Todd and J. S. Bard took over the small writing instrument firm with the intention of making gold nibs and fountain pens.

More than thirty patents on improvements to nibs and pens between 1878 and 1902 led to sales success, both in America and in the most important export market, Great Britain. The founders of the firm faced the problem of successors. They decided to sell a large part of their company to the London businessman, A. K. Watts. At first they wanted to limit the business to importing and selling American writing instruments, but the distance between the two continents soon made production in Britain necessary.

The Swan "Standard" insertion filler of black hard rubber, optionally available with a stainless steel or 14-karat gold nib and with ink passage over or under the nib, developed into a sales success as of about 1915. The Swan "Regular," almost identical to the models produced in America, was produced until well into the 1930s. The "Regular" standard filler and, as of 1912, the "Safety" were available in a wide range of variations, with gilded

The 1931 Swan line.

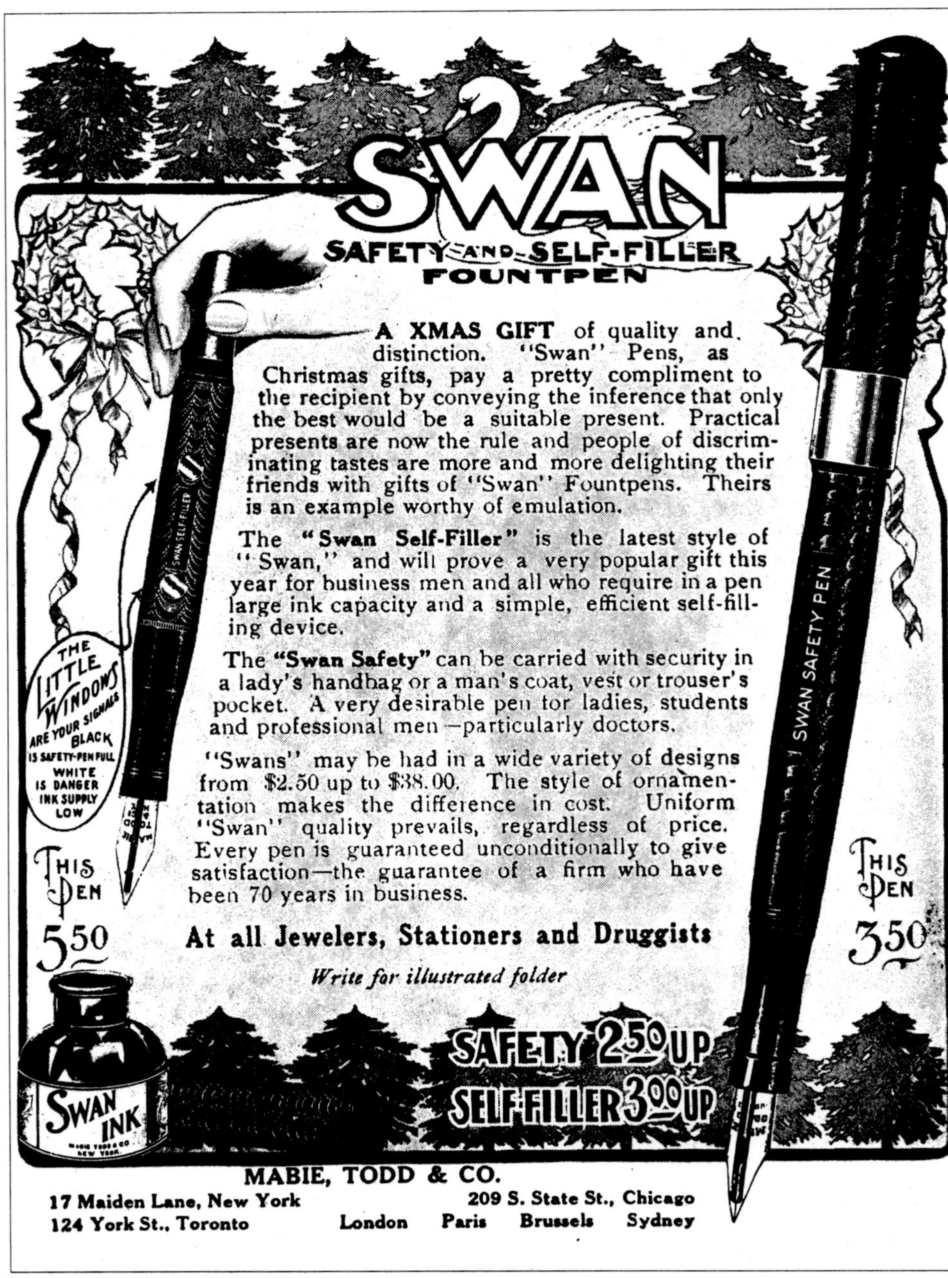

Swan pipette fillers of 1913.

decorative rings or completely covered with precious metals. The line of pens was accompanied by ink bottles and mechanical pencils that went by the name of *Fyne Poynt* (1920), *Mabie Magazine* (1916), and *Dace* (1910).

A special item of 1913 was the first Swan *Inkview* pen. Small glass lenses, set opposite each other in the sides of the barrel, allowed the user to check the ink supply in the reservoir at any time. The ink level was reflected by a thin strip of opal glass fastened vertically inside the barrel across from the sight lenses. Along with a simple eyedropper mechanism, these advantages helped Swan to attain great significance in Britain, North America, and many countries in Western Europe.

With some delay, the conservative Swan management followed the trend of the time and included the popular tube filler with a lever mechanism. Remarkably, newly developed insertion fillers appeared at the same time, as did the *Blackbird,* widely known as a low-priced pen.

In 1933 Mabie, Todd & Co. brought out a surprising new pen, the *Swan Leverless.* This tube filler without the familiar lever on the side was filled by giving a knob on the end of the barrel two half-turns. Inside the barrel, a metal rod was pressed concentrically against the rubber tube with every half-turn. When the knob was turned back, the pen filled with ink.

In 1936, Swan increased its sales success with an additional low-priced series below the *Blackbird,* the *Jackdaw,* and the newly developed *Visofil* filling mechanism which, similarly to the Parker *Vacumatic,* made possible an enlarged ink reservoir as well as a view of the ink level. A forerunner of the *Visofil* mechanism still had a piston rod. *Visofils* were available in a wide range of various colors and types of gilded cap rings.

In 1938 the Swan lever-filling pens and *Fyne Poynt* mechanical pencils were made with snake skin surface patterns and the leverless models with a lizard skin decor.

The outbreak of World War II caused the production of writing instruments to come to a stop for the time being. Instead of nibs and barrels, war materials had to be produced. And being a traditional English fountain pen manufacturer, Mabie, Todd & Co. was hit hard by the success of the new ballpoint pen. In 1951 the firm got into serious financial difficulties. Large portions of the company had to be turned over to the firm's ballpoint pen competitor, Biro Pens Ltd., for which it had already produced barrels and caps.

The new firm did business as Biro-Swan as of 1952 and produced the *Calligraph, Oxford, Warwick, York, Blackbird* and *Safety* models until 1957. Then the tradition-rich name of Swan disappeared completely from the fountain pen business. Promising new models such as the *Golden Swan* did not go into production. Twenty years later, Biro-Swan had become Biro-Bic.

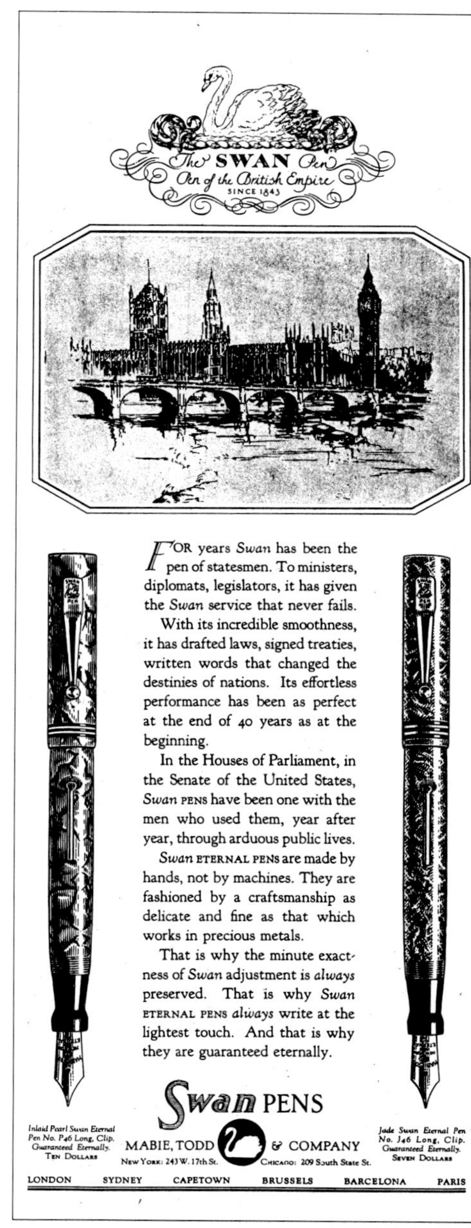

A 1929 advertisement.

Decoration

Just as architecture and art, furniture and clothing can, one of the most personal of all useful utensils, the writing instrument, can show a cultural history of designs. The American fountain pens, among others, provide particularly noteworthy examples.

Lavish Decoration (1860 to 1920)

The first fountain pens still more or less resembled their simple forerunner, the steel pen. A Soennecken, Waterman, or Parker of about 1890 had a pointed barrel end like the handle of a paintbrush. The cap had to be put aside when it had been removed. The fancy style that prevailed in the last years of the Nineteenth Century soon took possession of these simple black hard-rubber objects. Romantically designed metal bands with stylized flowers, spiral-shaped barrels or those with six, eight, or twelve side panels were regarded as chic and luxurious. Elements of Youth Style were mixed in, especially in the popular "filigrees"—pierced gold and silver

A desirable collector's item: the Parker "Lucky Curve" of 1907, with a sterling silver snake motif running around it.

panels. Whoever just wanted to spend a little money made do with waved or zigzag engraving.

Externally, the most varied products resembled each other. Variations were to be found only in the ink passages and filling systems. The most beautiful and valuable model of the time, as far as present-day collectors are concerned, was the Parker *Lucky Curve No. 37*, with a sterling silver or 18-karat gold snake running around it.

Elements of Victorian and Youth Style: fountain pen shapes of 1904.

Clear Art Deco Form (1920 to 1929)

The somewhat overdone added decorations of the fountain pen's early years gave way at the beginning of the Roaring Twenties to a new objectivity: the trend toward stylish writing. Two significant models set the prevailing fountain pen style of the time, the oversize Parker *Duofold* and the transition from black to colored hard rubber and celluloid-type materials. The typical Art Deco fountain pen had a cylindrical cap with a flat end, the clip usually ended in a ball, the barrel flowed into a trumpet-like holder with screw threads. For decoration, these pens were

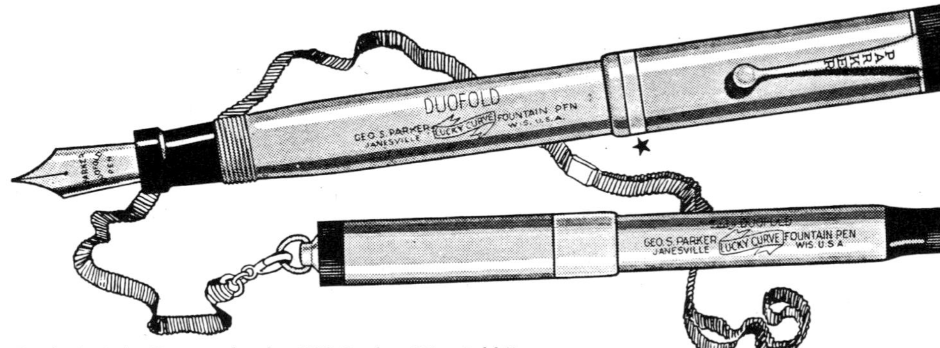

Typical of Art-Deco style: the 1923 Parker "Duofold."

adorned with cap rings and a gilded lever set in the shaft, often in a metal well. Above all, the Art Deco pens glowed in marbled colors of all shades. The most important representatives of this era in America, in addition to the Parker *Duofold*, were the Waterman *Ripple* and *Patrician* and Sheaffer's *Lifetime Pen*.

Streamline Form (1929 to 1939)

At the end of the decade, Sheaffer opened a new era in the history of fountain pen style with the streamline or torpedo form. The new generation went by the name of *Balance* and featured a shaft and cap running to points. To the user, this suggested balanced weight and smooth, light writing. Sheaffer had obviously complied with popular taste here. Automobiles, radios and lamps were also appearing in smooth aerodynamic forms. Parker and Eversharp followed the trend. With the *Doric,* Eversharp produced one of the most harmonious, original and beautiful of fountain pen designs.

Sheaffer's "Balance Pen."

Modern Styles (1939 to 1970)

The 1940s were typified in America by technical progress, record sales, and stylistic triumphs. Parker invested about a quarter million dollars in a new fountain pen that was to prevail for the next thirty years, the *Parker 51*. Designer Laszlo Moholy-Nagy created a slim, simple, yet elegant pen. Its design must have seemed to the people at that time like a preview of 1951 styling. The crude cap of thick plastic gave way to a slim metal cap. The once large nib was reduced to a writing tip almost completely surrounded by plastic. Today the *Parker 51* is regarded as one of the most successful and most often-copied pens of all time. Its trailblazing form subsequently influenced pens made in America, Japan, and Europe.

The Super-Slim Types (since 1970)

The function is the basis of the design. This rule of Bauhaus adherents was taken to its limit in the 1970s with the pen-point. The representative pens of the modern era presented themselves as elegant and restrained cylinders without snorkels. The Italian Aurora firm's *Hastil* model ranked among the pioneers, but so did haute-couture brands such as Dunhill, DuPont, Nazareno Gabrieli, as well as Montblanc, Pelikan, Lamy and even the classic among the slim-line types, Cross. The favorite material was stainless steel or layered brass, for only metal allowed the required small diameters and thin walls.

High tech was the slogan of the progressives, a concept that appeared in the audio-video world, in home furnishings, and computer designs as well as writing

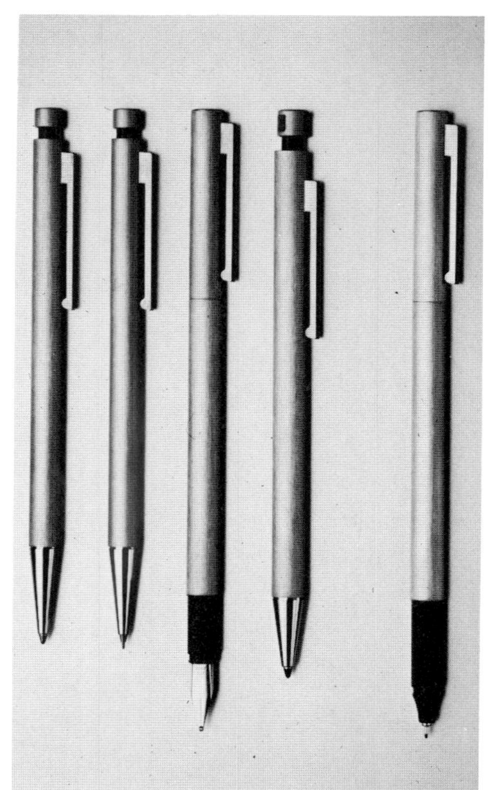

Slim and simple: The Lamy cp 1 is a typical representative of modern writing instrument styling.

instruments. The preference of people with futuristic tastes coexisted with that of those who looked backward. The style of the 1950s enjoyed a splendid comeback as a contrast. The Pelikan *Souverän* piston filler, Montblanc *Meisterstück*, Parker *Duofold* and pens by Waterman, Omas and Sheaffer are regarded today as classics among the world's desirable pens.

Caps and Clips

It is the cap's job to protect the sensitive nib from damage and drying. The early caps were relatively short and only covered the nib and a little of the grip when in place. Parker was the first, in 1898, to design the front section and the cap so that the cap passed well over the grip and could be pushed onto the barrel, achieving a better fit. Two years later, a further improvement followed. The external cap was lined with an inner cap for additional sealing. To this day a lining of flexible plastic serves as a sign of quality, indicating hermetic sealing. With the introduction of the safety pen, the manufacturers switched from slip-on to screw-on caps. It was more than thirty years later that the convenience of the slip-on cap was rediscovered.

The clip too, the pen's nose, so to speak, gained recognition as a separate part of the pen's equipment only late. In the 1920s people were still buying the fountain pen and the clip separately. The slip-on clip generally had a half-ring that was pushed over the cap and held itself in place. Swan offered a sheet-metal case instead of a clip to hold its clipless pens in a trouser or jacket pocket. Waterman was the first, in 1906, to stress caps with clips. The "Clip-Cap" had a wide clip with two rivets holding it permanently to the cap, "straight from the factory." Parker followed in 1907 with a patented "lever-locking clip," a short metal clip bonded to the surface of the cap. When one slid the cap onto the front section to close the pen, the clip was pushed out and could fulfill its clipping function in the vest or jacket pocket. For the *Endura,* Conklin used the spring power of the clip fastener for easier insertion in the pocket. When one pushed on the upper part of the clip, it spread out from the cap and could be slid easily over the edge of the pocket.

Until the beginning of the 1920s, the clip was located approximately in the middle of the cap. Then the Art Deco styling principles demanded stricter forms. Thus the clip attachment moved close to the end of the cap, which in turn allowed a deeper and thus safer insertion of the pen.

Fountain pen clips all had approximately the same form until the 1930s. The rod always ended in a ball to protect the clothing from damage. Then someone got the idea of using the clip as an identifying mark of the brand. Wahl-Eversharp turned its clip into a fork holding a movable ball. Parker made the arrow its characteristic clip form, Pelikan made its clip in the shape of a pelican's bill, and Waterman used a V-shaped clip.

Pens with clips were preferred by men, but women scarcely needed them, as they did not carry a pen in their jacket or blouse pocket. A fine necklace was preferred, to which a pen could be hooked, always ready to be taken in hand and used. A number of manufacturers produced their fountain pens and mechanical pencils with a ring instead of a clip at the end of the cap. This in turn fascinated the men. Fountain pens and pencils were worn on their watch chains as "vest-pocket writers."

At the beginning of our century one bought the pen and the clip separately.

The practical sheet metal case replaced the clip for early Swan fountain pens.

Fountain pen clips were and are made of steel or springy sheet metal, formed by special tools. The material, shape, and type of fastening (ring, box, or crimped) give it its springiness. Another generation of clips were stiff and made of massive steel. They derive their mobility and holding power from a coil spring hidden inside the cap. Sheaffer used such a solid clip in 1959 for the PFM model. Today Lamy uses this kind of clip on many different models.

The familiarity and attraction of the major fountain pen brands tempted many unknown manufacturers to try to exploit them. Imitations of Parker clips were made in America, Italy and the Far East. Even today, a government-owned firm in Communist China makes fountain pens with clips that can scarcely be distinguished from a genuine Parker clip. In the 1950s, Pelikan counted more than 35 imitations of its typical pelican-bill clip. For decades it was regarded as chic to display the clip of a brand-name fountain pen on one's jacket or shirt pocket. For numerous manufacturers, the export of caps to Africa or the Far East, ostensibly as replacement parts, is said to have exceeded the sale of complete pens at times. Wearing the status symbol on their shirt pocket must have seemed more important to many people than owning a complete fountain pen.

Soennecken's "gold fountain pen carrier prevented the pen from falling out of the pocket" in 1906.

The vertical models show the Parker clip recessed in the cap and automatically pushed out when the pen was closed; 1914.

Conway Stewart

This old English firm owes its name to chance. Its founders in 1905 were Thomas H. Garner and Frank Jarvis. When Mr. Garner had to travel incognito to Scotland one day, he used the name of a famous comic duo who regularly appeared in the Collins Music Hall at Islington at that time. And because the colorful fountain pens reminded him of the brightly clothed performers, not only the line of writing instruments but the entire firm was given the stage names of the two comedians: Conway Stewart.

Conway Stewart specialized in low- to medium-priced fountain pens made of hard rubber, and later of galalite and celluloid with lever filling systems, under the names of *Conway, Duro, Dinkie, Dinkie Major, Dandy, Scribe, International* and *Universal*. The small-size *Dinkie* for ladies was one of the most successful and renowned Conway Stewart products. It was put on the market in 1932.

All the firm's writing instruments were produced and assembled by more than four hundred employees. By the end of the 1930s, more than a million fountain pens had left the Conway Stewart works in London.

After World War II the market demanded a practical conversion to modern injection molding. Production of the beloved *Dinkie* in colorful marbled shades had to be halted. Almost one-third of the firm's production went to the Woolworth chain under the name of *Winfield*. In 1975 the firm, ranking among the pioneers of fountain pen production, had to cease production and liquidate.

In 1983 Maltgate Ltd. bought the production facilities and the brand name and began to manufacture writing instruments anew in a modern factory in Pontypool.

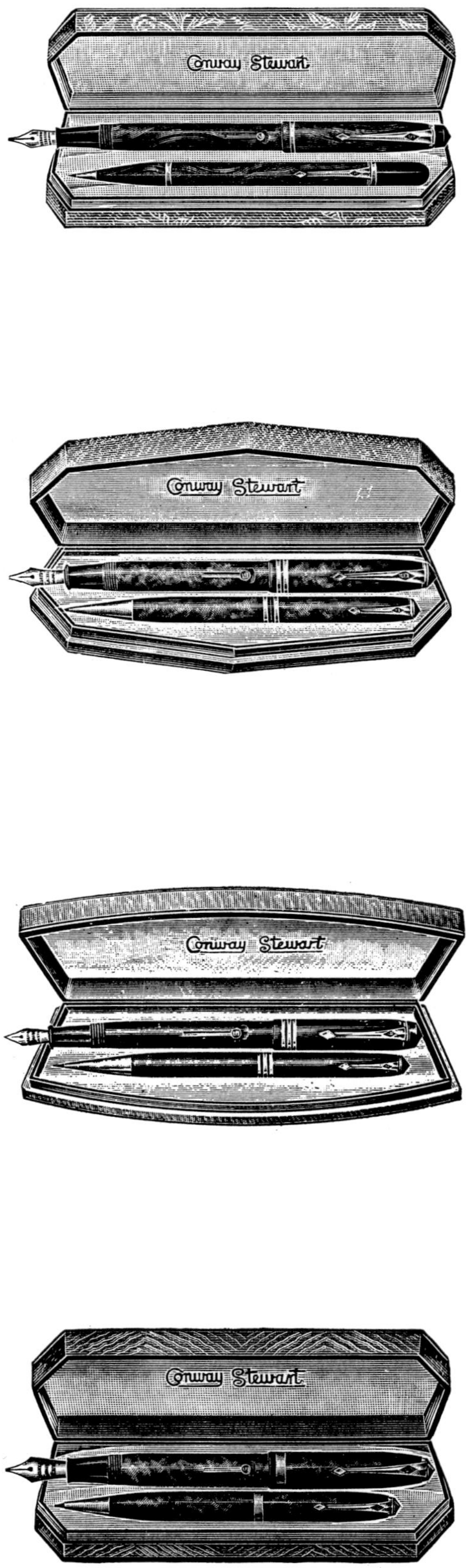

Gift cases of 1938.

A 1938 price list.

A 1932 advertisement. ▷

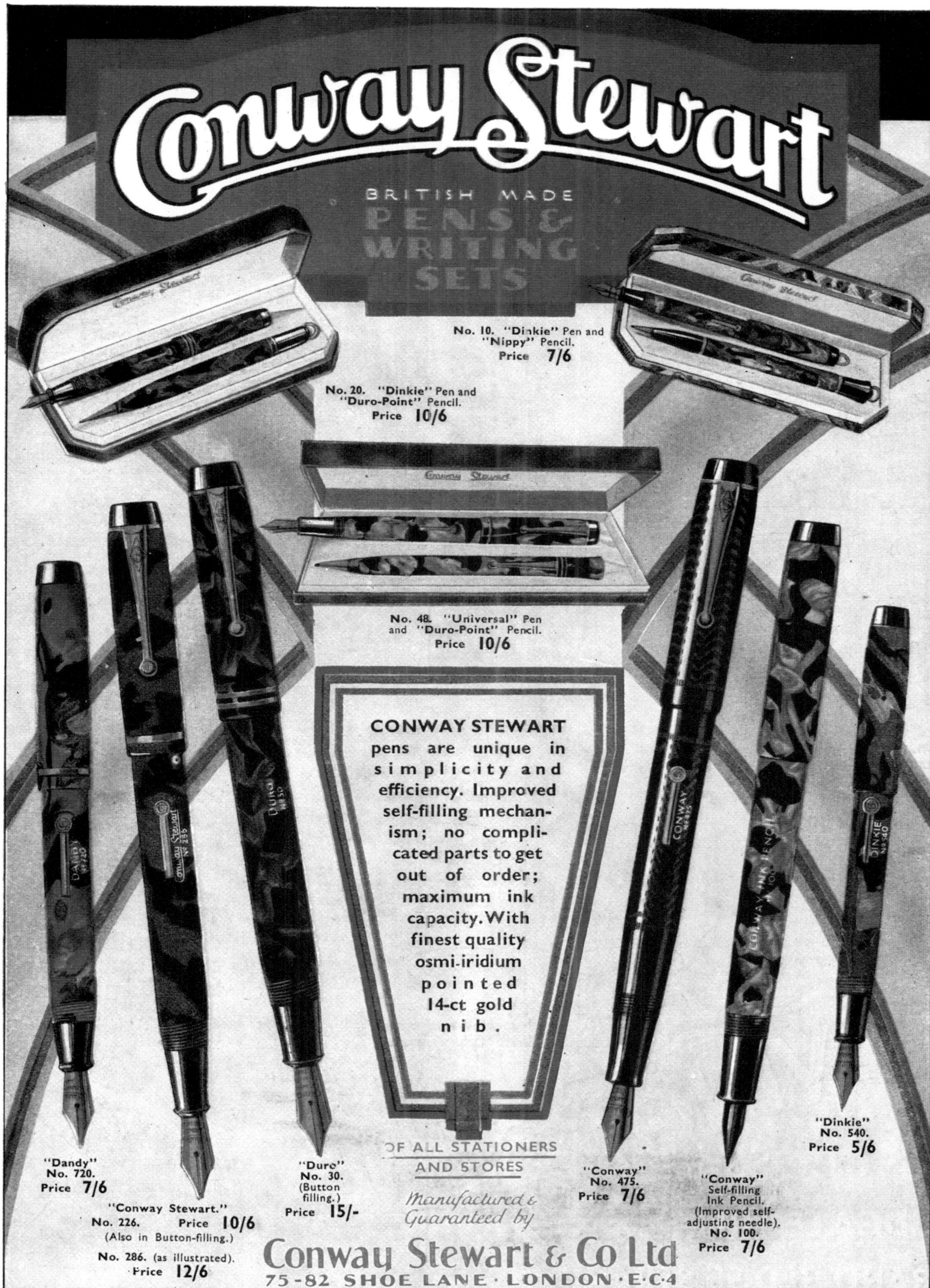

Conway Stewart
BRITISH MADE
PENS & WRITING SETS
No. 10. "Dinkie" Pen and "Nippy" Pencil. Price 7/6
No. 20. "Dinkie" Pen and "Duro-Point" Pencil. Price 10/6
No. 48. "Universal" Pen and "Duro-Point" Pencil. Price 10/6
CONWAY STEWART pens are unique in simplicity and efficiency. Improved self-filling mechanism; no complicated parts to get out of order; maximum ink capacity. With finest quality osmi-iridium pointed 14-ct gold nib.
OF ALL STATIONERS AND STORES
"Dandy" No. 720. Price 7/6
"Conway Stewart." No. 226. Price 10/6 (Also in Button-filling.)
No. 286. (as illustrated). Price 12/6
"Duro" No. 30. (Button filling.) Price 15/-
"Conway" No. 475. Price 7/6
"Conway" Self-filling Ink Pencil. (Improved self-adjusting needle). No. 100. Price 7/6
"Dinkie" No. 540. Price 5/6
Manufactured & Guaranteed by
Conway Stewart & Co Ltd
75-82 SHOE LANE · LONDON · E·C·4

Industrial Report:

A Fountain Pen is Made

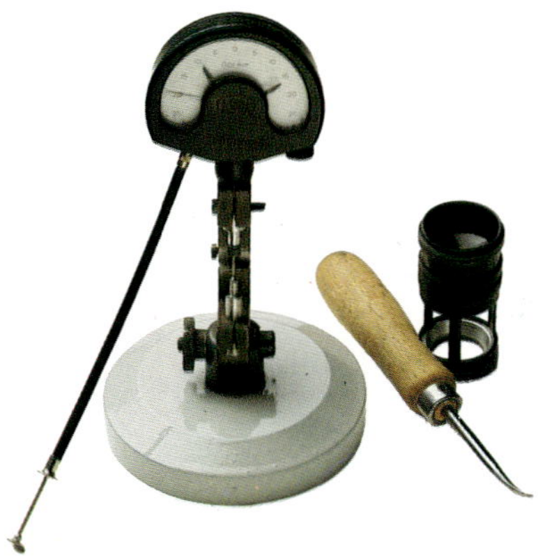

Builders and designers confer on technical construction and design. The construction plans take form on the drawing board.

Preliminary work on the production of special injection molding machines.

Precision to 1/100 of a millimeter is necessary for tool production.

Cellulose acetate is the raw material used for making plastic parts, rather than the easily inflammable celluloid used in past decades.

Forming the connectors by hand in hot-water baths.

A diamond cuts a gentle contour in the cylindrical basic form.

Part after part is carefully tested.

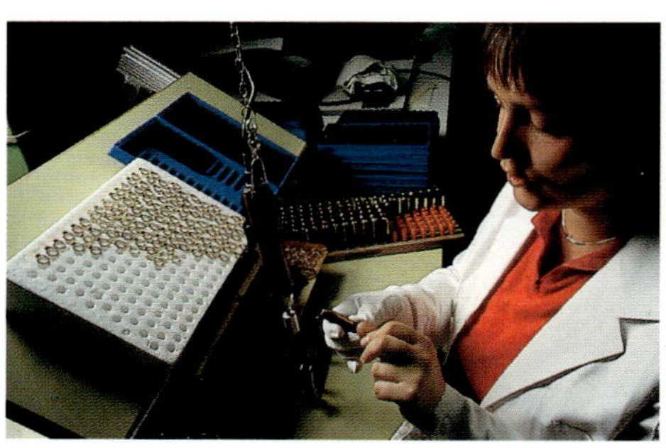

Assembling the cap by hand.

Completely assembled caps.

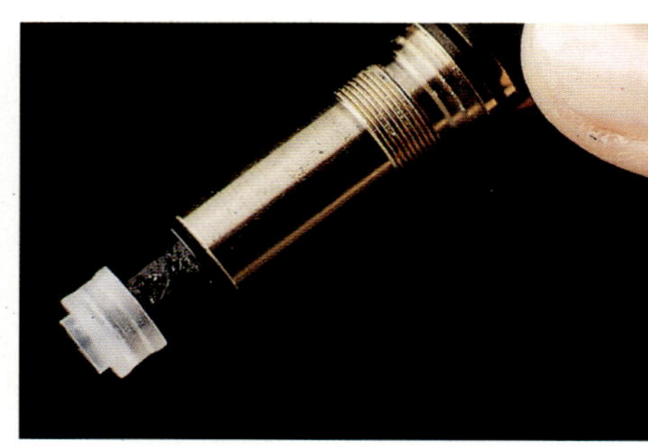

Installing the piston mechanism.

Every component is given a rigorous test before it can be worked further.

Piece after piece is brought to a high polish on the polishing disc.

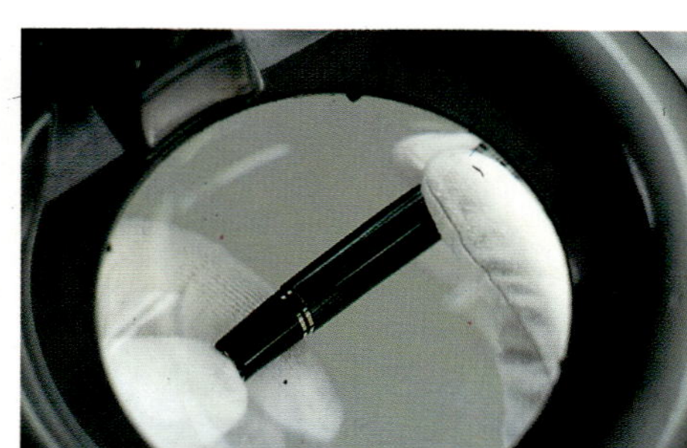

A careful look through the loupe. From now on, the finished fountain pen is handled only with white gloves.

Industrial know-how, precise handwork, intensive research and careful choice of materials are the basis of the manufacturing of that precision instrument, the fountain pen. Up to 100 work processes and more than 30 days are needed to make a high-priced fountain pen.

The tested pens draw exact lines on paper drums during ink-flow testing.

Hardness test at 40 degrees below zero Celsius . . .

. . . and at 70 degrees Celsius in the heat chamber.

Test and test again. Only faultless pens bear the seal . . .

"Made in Germany."

The Ballpoint Pen

"Success has many fathers."
Proverb

LOUD 1888

LAMBERT 1891

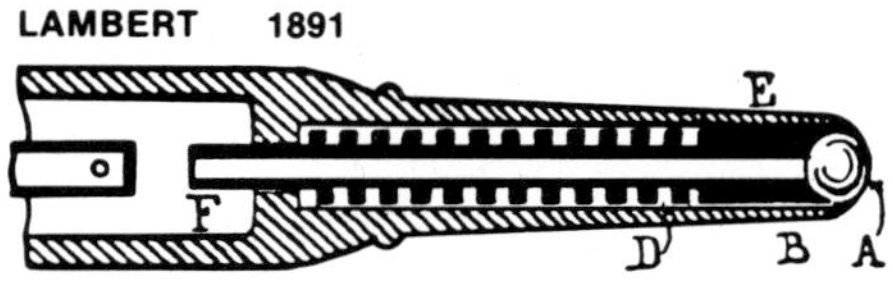

The idea of writing with a ball is over 100 years old.

To whom does the world owe thanks for the invention of the ballpoint pen? That is in dispute. It has occurred to numerous inventors that the ball is superbly qualified to be a rolling writing element.

The main problem that ballpoint pioneers like the Americans J. J. Loud, in 1888, and Mr. Lambert, in 1891, could not solve was the ink. Loud was thinking not so much of a writing pen as of a marking pen for crates, packages, and other uneven surfaces. According to his patent, a coil spring in a cylindrical tube pushed a piston against two balls which in turn pressed the writing ball against the mouth of the writing tube. The pressure of writing allowed a small gap to open between the writing ball and the mouth, through which the writing ball could be supplied with flowing ink. After being used, the piston of Loud's writing instrument was to be screwed tight so the ink would not run out unintentionally. Such early inventors got only as far as the patent. None of these models went into production. Obviously the consistency of the ink available around 1890 was not sufficient to cover the writing ball consistently.

In 1929 a ballpoint pen called *Exact* was displayed at the Leipzig Fair as the newest advance in the field. The *Exact* had to be filled with a tube. A special screw key served to control the ball.

Other inventors followed. One of them was the Hungarian Lászlo Biró. Born in 1899, this painter, engineer, and member of the Hungarian Royal Academy of Science already had some thirty small inventions and patents to his credit, hand-operated washing machines, an unbreakable lock, heat-resistant tiles and the like, when he and his brother Georg registered the first Hungarian idea for a ballpoint pen in 1938. Earlier in life, as an officer in World War I, he had told his colleagues of his ideas for improving the fountain pen. During the 1930s Biró published a cultural magazine and could not help but encounter the problems of printing, particularly the quick-drying printing inks used at that time. He quickly got the idea of using these quick-drying inks for writing. The soft paste required a completely different writing instrument than the traditional pen. Nor did a piston filling system seem suitable. Hungary's allegiance with Germany in World War II forced the politically involved Biró brothers to leave Europe in 1940. They found a home in Argentina, whose president, August Justo Lázlo, had met Biró during a vacation in Yugoslavia. Justo had noticed Biró writing a postcard in a hotel lobby with a prototype ballpoint pen.

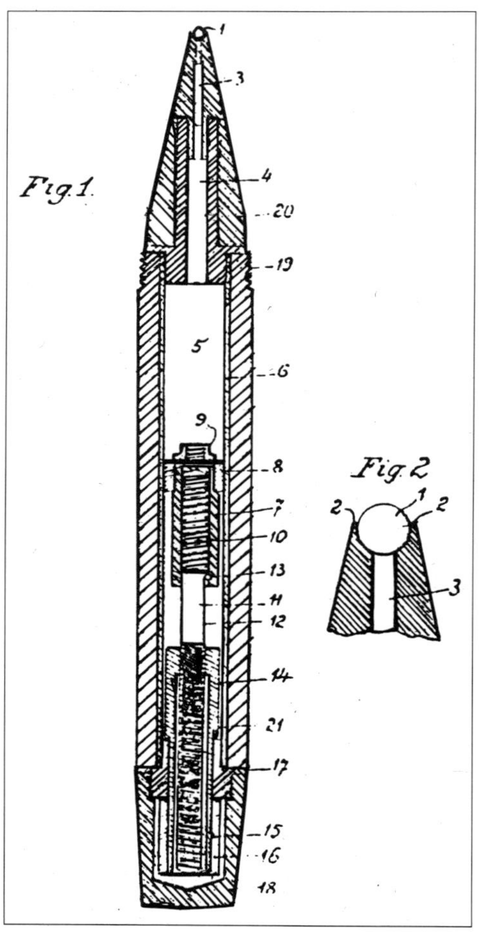

The original Biró patent of 1938 describes a refillable ballpoint pen with turning piston. Later Biró converted to the replaceable cartridge system.

For two years the Biró brothers experimented with thousands of formulas until they found an oil-base paste with the ideal viscosity and capillarity to supply the writing ball steadily, without drying out or leaking out. The small steel ball was pressed into a brass mount.

In 1943 they and others founded a company to make commercial use of the patent for an improved ballpoint pen that was granted them in the same year. As refugees without capital, the inventors brought only their patent as their contribution to the firm of Eterpen S. A. As a result, they received only one third of the sum of two million U.S. dollars that Eterpen was able to earn in 1944 for the sale of the North American license rights to Eberhard Faber and Wahl-Eversharp.

Around 1944 an English businessman, Henry George Martin, bought into the Argentine firm of Eterpen S. A. and obtained the sole patent rights to the Biró invention. Martin quickly realized what kind of potential the new writing instrument had. For a long time, the Royal Air Force pilots had been bothered by fountain pens that leaked at high altitudes. Martin offered the U.S. Air Force and the RAF the new pen for testing. The pilots were enthusiastic. Before 1944 ended,

Martin's "Miles Aircraft Co. Ltd." delivered more than 30,000 of the new ballpoint pens to the British Air Ministry. This was a remarkable accomplishment, for it took more than six weeks to polish the little balls between two counter-rotating steel plates until their uneven spots had been reduced to less than 1/100 the thickness of a human hair.

While use of the "Biró" in England was limited to the cockpit crews of the RAF warplanes, Eterpen put the sensational new pen on the open market in Buenos Aires in 1945—the *Birome* cost the goodly sum of 25 pounds. At the same time, the license holders, Faber and Wahl-Eversharp, began to sell the pens in North America under the name of *Capillary Action* (CA).

At the last minute, fate played a trick on the Biró brothers. The American industrialist Milton Reynolds went into competition with them by inventing his own refillable ballpoint pen. How he managed to circumvent the Biró patent is not known. The world watched anxiously. The practical-minded American public developed tremendous enthusiasm for the *Satinflo* that, according to a lot of advertising, could write just as well in Arctic ice or under water as in the stratosphere without blotting or drying up. "Write for a year or for 275 hours without refilling," Reynolds promised the astonished public. And when it was time to refill, Milton Reynolds would take care of this small matter if one sent him the pen and fifty cents.

The Reynolds ballpoint pen, though, was not a thing of beauty. The dull-looking object consisted of an aluminum housing with cap and clip. The writing point was a thin tube with a steel ball, about one millimeter thick, bedded in it and turning freely in its setting. The writing ball was supplied with writing paste via four to six hair-thin channels. Unlike fountain pen ink, the ballpoint paste did not sink into the paper but only stuck to the surface. A result was a lack of smear resistance.

Reynolds' famous pen was first sold in October of 1945 in the famous Gimbels department store in New York, at the remarkably high price of $12.50, and was grabbed out of the sales clerks' hands. In the first days of a continuing sales campaign, the stationery department of Gimbels alone took in $100,000.

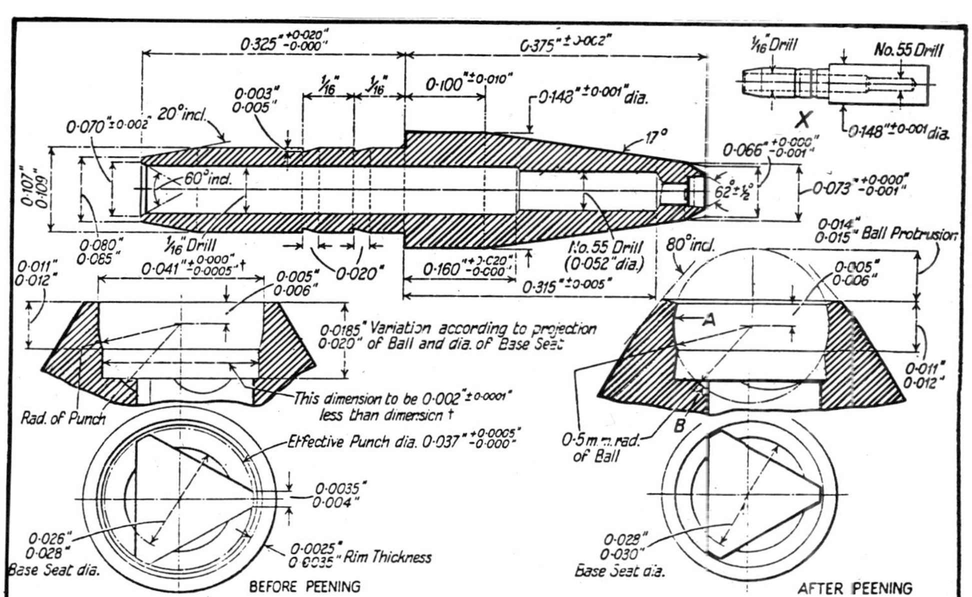

Excerpt from a technical drawing for the construction of the Biró cartridge point. The manufacturing tolerances required a precision of approximately 1/100 of a millimeter.

Other companies realized that the new ballpoint pen could have a great future. In 1946, Eversharp was followed by a whole host of smaller manufacturers with their own ballpoint creations. Papermate took the lead in the ballpoint pen market by the early 1950s. The heavy competition caused prices to drop. From the original $12.50 they soon fell to fifty cents. Though the public treasured this new instrument that wrote for months without refilling, the quality of the first ballpoint pens was wretched. The heavy ink paste crept only slowly over the rotating steel ball. Blanks and blots were the results. Some manufacturers experimented with the familiar rubber tube of the fountain pen as a paste container, while others used brass tubes.

Only in 1954 did Parker bring out its first ballpoint pen, the famous *Jotter*. The relatively long development time had been used well by Parker. Unlike other ballpoint pens, the *Jotter* avoided one-sided wear on the ball mount, designing it so that the cartridge turned ninety degrees every time it was engaged. A further improvement followed in 1957. Instead of the customary steel ball, Parker now used a ball of tungsten carbide with a surface that was not smooth, but was made with about 50,000 tiny surfaces and raised areas, comparable to a golf ball. This new type of ball moved firmly on all kinds of surfaces, whether rough or smooth. The *T(extured)-Ball Jotter* became a worldwide bestseller, and over seventeen million of them were sold per year.

Eberhard Faber and Eversharp took over the development of Biró's invention in North America.

In Germany, pencil manufacturers such as A. W. Faber, plus smaller firms and branches, were the first to take up production of ballpoint pens. Metalworking shops in the Schwarzwald and Odenwald developed into large firms—the good Schneider cartridge is as an example of what many achieved.

Next to the wooden pencil, the ballpoint pen is probably the most frequently used of all writing instruments today. The yearly worldwide production is estimated at more than a billion pieces. The production methods have been developed almost to perfection, and the resulting high quality has made the ballpoint pen a problem-free writing instrument for everybody. Improvements scarcely seem possible any more, and yet one firm, Fisher Price Pen of the USA, was able to move ballpoint pen production another leap forward. The ballpoint pen's tendency to stop writing when used "overhead" or in the emptiness of space gave NASA, the American space agency, a big problem. The astronauts had to be able to keep a written logbook. Mr. Fisher came up with the right idea. He developed the *Space Pen,* a ballpoint cartridge that used gas pressure of about 5 bars to push the paste toward the writing ball. Thus he offered a pen that did its job willingly, independently of external influences, at temperatures from -50 to +200 degrees, under water, and even in the vacuum of space.

Scarcely any other product has been so exploited by the advertising and souvenir industries as the ballpoint pen. From an engraved or printed advertising text, or a lady who unveils herself as the pen is turned, to the shape of a banana or tennis racket, or an overly long type that can be tied in a knot, everything imaginable has already been done and will continue to be done in the future in countless variations.

The last step in ballpoint development is the ink roller—a Japanese invention that combines the advantages of the ballpoint and felt-tip pens. Instead of a paste, liquid ink is put on the paper by the "roller ball"—producing a script that at first glance looks very similar to that of a fountain pen.

Scarcely anyone still remembers the inventor of the ballpoint pen, Lászlo Jozsef Biró, who called himself Ladislo Biró in Argentina, his second homeland. Like many other inventors, he did not succeed in gaining much from the economic success of his invention. Paid off with a relatively laughable sum, he left Eterpen in 1947 and turned to surrealistic painting, with which he had already had some success in Budapest. Some of his works are still to be seen in the Budapest Art Museum. In 1969, as the advisor to an Argentine ballpoint pen firm, he came into contact with his invention again, and later he worked on uranium enrichment for the Argentine Atomic Energy Commission. The man who, through his invention, probably had more influence on written communication than anyone else died on October 24, 1985. In many lands his name came to be synonymous with the ballpoint pen—which is known simply as "Biro."

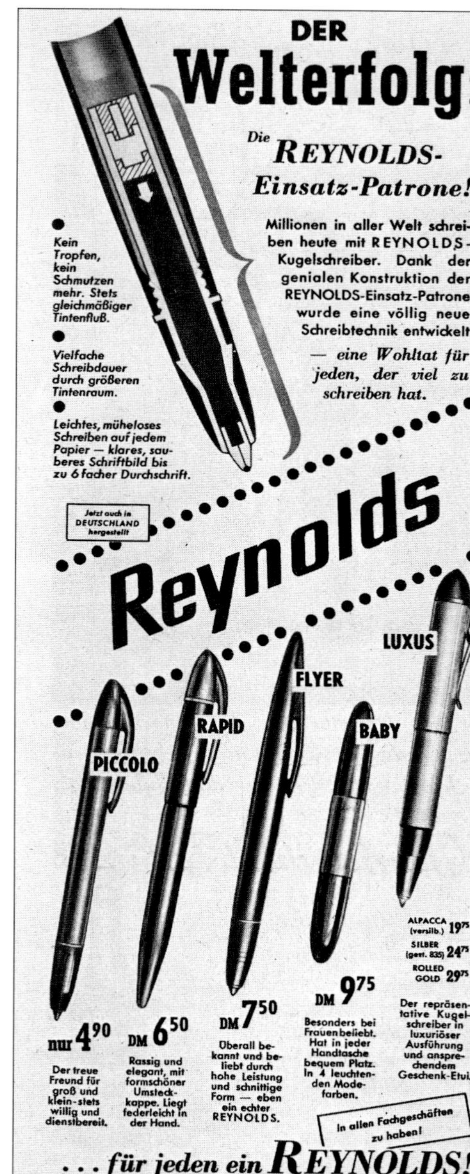

The first Reynolds ballpoint pens were torn out of the salespeople's hands in December of 1946. A 1949 advertisement.

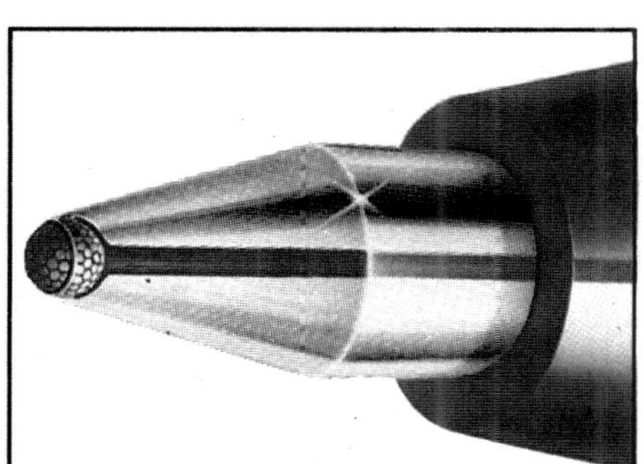

A modern ink roller with a ceramic ball.

The Technology of the Ballpoint Pen

The cartridge determines the quality of a ballpoint pen. It consists of a brass, copper, nickel, or plastic tube ending in a tip of German silver, steel, or brass. Manufacturing these tips demands extreme precision. Four or five very narrow channels lead the writing paste to a bowl-shaped ball bed, into which the completely round steel, tungsten carbide, or stone (ruby or silicon carbide) ball, about 0.8 to 1 mm in diameter, is pressed. If it is positioned exactly, the ball, when at rest, closes the cartridge and thus prevents drying. The lifetime and writing quality of a ballpoint pen cartridge depend on the ball and its exact position in its bed, with tolerances of only thousandths of a millimeter, as well as by the kind of handling they get from the user. Damage to the tip by falling or being hit will unavoidably lead to spotting or leaking of the writing fluid.

When used for writing in a vertical (point up) or strongly angled position on walls or other vertical surfaces, air can get into the tip of the cartridge. It forms bubbles in the paste, and the column of paste gradually slides toward the end of the cartridge. As a result, the cartridge can be persuaded to write only with difficulty, if at all. To prevent the paste from running out, parts made of metal, plastic (silicon), or foam are installed, equipped with small air channels for the necessary air supply. In large-size cartridges, so-called gliding plugs keep the paste moving smoothly.

Standard dimensions and paste formulas have been set in Germany by DIN 16554 norms. Accordingly, solutions of high-grade pigments must be used that are "document true," resistant to falsifications such as erasure. Only cartridges made to DIN standards may be used in ministries, public offices and banks. DIN cartridges must also contain a paste that will dry to be smear-proof within twenty seconds.

The most frequently used cartridges are:
A 3 = cartridges for ballpoint pens (3 mm)
B 3 = cartridges for ballpoint pens (4.1 mm)
D 1 = cartridges for multi-color ballpoint pens
G 1 = oversize cartridges

Ink rollers, like ballpoint pens, have a stainless steel ball bedded in a mount of stainless steel or nylon. Unlike the paste ballpoint pen, the ink roller, thanks to its liquid ink, cannot smear.

With all its handiness and simplicity of use and functioning, the ballpoint pen nevertheless has one decisive disadvantage compared to the fountain pen. The position in which a ballpoint pen writes best is not the natural writing position. Steady control and even writing result only when the pen is held at a very steep angle, which is uncomfortable and easily leads to cramping and tiring. These are surely not good conditions for bringing out and developing individual handwriting! Perhaps this is also a reason why the ballpoint pen is banned in the lower grades of schools in several European countries.

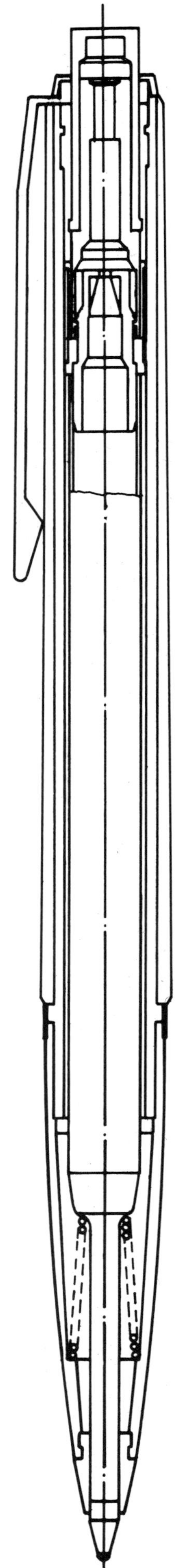

Cross-section of a modern push-button ballpoint pen.

The typical Papermate push-button ballpoint pen, from 1963.

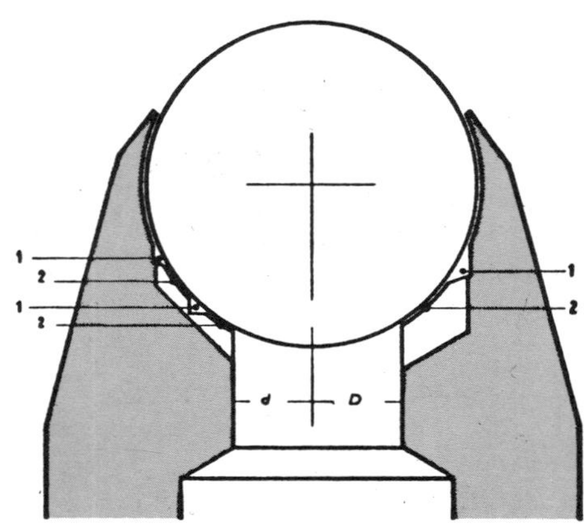

Cross-section of a ballpoint pen point.

1. Feeder ring canal
2. Support surface
d/D. Paste-feed channel.

Imagine a Pen

not much longer than your cigarette!

. . . when open it's a full size pen—with full ink supply!

This Kimberly Pockette is a new EVERSHARP writing wonder! When closed—it's little longer than a cigarette. Drop it in any pocket, in vest, coat, trousers. It won't leak—it can't. It *rolls* ink on *dry* . . . writes on and on and on . . . with the Magic Sphere Point so precision-fashioned that you can write *at any angle!*

You'll want this Kimberly Pockette, no matter how many pens you own! It's so convenient, so amazingly smooth-writing. Handsomely styled in your choice of mannish solid colors.

With 14-K gold-filled band, only $7.50 . . . (tax included)
With Chrome band, only $4.95 (no tax). Buy yours now.

Kimberly
POCKETTE
by EVERSHARP

P. S. *Buy two—so your wife won't "borrow" yours. Get her a Kimberly Pockette for her handbag . . . in feminine high-style colors.*

A "writing wonder"—no bigger than a cigarette—was the first pocket ballpoint pen of 1947.

A. T. Cross Company

America's oldest manufacturer of high-quality writing instruments was founded in Providence, Rhode Island, a center of the American jewelry industry comparable with Pforzheim, Germany, in 1846. Alonzo T. Cross, the son of British immigrants, originally worked at a trade that had nothing whatever to do with elegant, high-priced writing instruments—he was a trained cooper. In the New World in the middle of the nineteenth century there was a demand for jewelry of all kinds. Cross used his mechanical talents to work metals for jewelry. He made thermometer cases, cigar holders, lighters, and imaginatively decorated pencil cases. It may have occurred to him that if a wooden pencil, which becomes shorter and shorter from being sharpened, needs a lengthener, why should there not be a pencil that needs neither lengthening nor sharpening? No sooner said than done. A. T. Cross developed the first metal mechanism that allowed a pencil-lead to be moved out of a metal tube and back in again. The prototype of his invention was presented to the astonished American public in 1864. And it was a success. Twenty-seven patents, including one for the "Magic Telescopic Pencil" (1869), were to follow.

In 1879 Cross added the "ink pencil" to the mechanical pencil. A small ink reservoir in the shaft supplied the writing tip, consisting of a fine rounded metal tube, with ink. The expensive ink pencil may not have satisfied the public demand completely; it was given up after five years of production. After this experience, A. T. Cross concentrated completely on the pencil. The mechanical pencil introduced in 1879 allowed the use of a longer lead that could be moved out and in by turning a button. A gripper mechanism prevented the lead from sliding backward while in use. The mini-model was particularly preferred by ladies, the maxi-model by gentlemen.

Forty-six years went by before A. T. Cross made another writing instrument that used ink. The success of the fountain pen encouraged the addition of such a device to the Cross line. The finish of 12-carat rolled gold, the slim conical form and the longitudinal engravings on the barrel and cap are striking. They originated the typical Cross design that became recognized all over the world.

When the Cross firm celebrated its hundredth anniversary in 1946 by introducing the "Centennial Pencil," the ballpoint pen had been born. The tremendous demand made it possible for Cross to change its former company policy from the ground up. The small but classy firm intended from then on to concentrate exclusively on high-priced, elegant writing instruments, sheathed in costly precious metals, for customers of distinction. Mechanical pencils and ballpoint pens, both with the same mechanics, were offered in various designs. These instruments, made with the greatest precision (and unchanged externally since 1946), indicate their worth by the size of the black ring at the upper end. The smaller this ring, the more valuable the surface material. Only in 1982 did the A. T. Cross Company extend its assortment once again with a fountain pen.

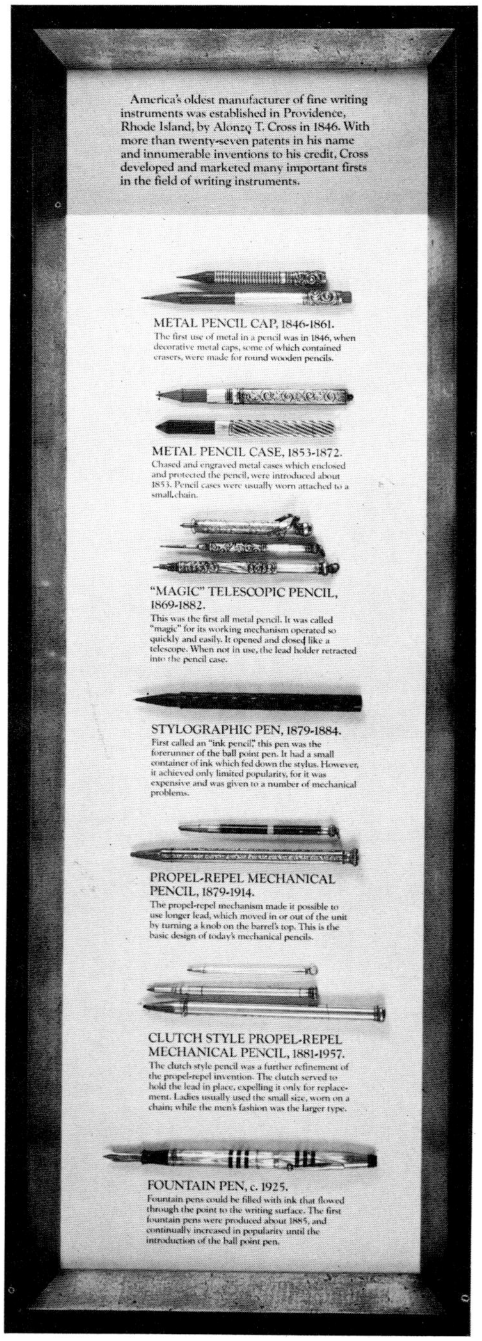

From the pencil cap to the fountain pen. The fountain pen shape of 1925 was subsequently used for all later Cross writing instruments.

How a Ballpoint Pen is Designed—the Colani Example

Writing instruments are tubes, more or less long, with a writing point of some kind, and with pigment inside. The average person would probably define them in such terms. When the designer Luigi Colani was commissioned in 1979 to design a ballpoint pen, this man who describes himself as a philosopher of three-dimensional shapes, a sculptor, environmental physiologist, and expert in aerodynamics, already had a definite concept. This most imaginative creator of shapes had already proved by numerous examples that ordinary utensils can be turned into exclusive works of art. From the baby carriage to the automobile, glider, ship, telephone, and motorcycle, he has designed everything that modern people need in life or at least regard as necessary. Even new types of potty-chairs, cakes, eyeglasses, toothbrushes, perfumes, and wristwatches have sprung from the head of this remarkable architect of shapes.

"My model is nature," says this man, who has now reached the age of sixty. "Straight-line forms are unnatural. The fish in the water, plants, the parts of the human body have curved, often streamlined forms. Lines drawn with a ruler are artificial, drawn by human hands. Utensils that are to serve human beings have to have shapes that are adequate for them . . ."

Even in the first preliminary talk that the manufacturer had with him, it was clear: in spite of the strict requirements involved, such as technical uniformity, economic production potential, and a competitive selling price, a "Colani" had to be related to nature.

Equipped with a thick sketching pad, Luigi Colani withdrew into the solitude of his extensive studio, to emerge five weeks later with five dozen large-scale drawings of white on black cards. In long discussions, developmental engineers and production technicians, marketing specialists, and sales personnel chose the basic design that seemed most suitable for further development.

Building on the chosen basic concept, others followed, essentially much more detailed drawings, and then a first model. Made of plastic, brightly painted, it offered the first idea of how the pen that would go into production later would feel in the hand. The "innards" were still lacking. They would be designed by the developmental engineers. The three-sided grip section was to "correspond to the natural writing position of gripping the instrument with three fingers." The clip should no longer be an added-on separate part but, formed like the bill of a waterfowl, be a part of the total form. There was no place for the otherwise customary push-button to move the oversize cartridge into position. A solution was found by cooperation. A knobbed lever attached on the side fit into the design without any disadvantages.

Luigi Colani is regarded in Japan as the most outstanding European designer. From the orthopedically optimal chair to a streamlined motorcycle, his spacious studio in Bern contains hundreds of examples of his wealth of ideas.

Correctly shaped furniture for sitting.

Luigi Colani: *"Arriving at the natural writing position."*

The resulting model was given one last polishing by its designer. This was a job that requires an absolute feeling for form, for in a utensil like a writing instrument, radii and diameters of a tenth of a millimeter are often vital to the total effect.

Finally production of the "Colani No. 1" was ready to begin. Injection tools that had to be finished by hand with the highest precision were contracted for, production facilities for the assembly of the individual parts were prepared. Eighteenth months after the contract was given to Colani, the point had been reached at which the new ballpoint pen could be presented to the public. Today the "Colani No. 1" ranks among those ballpoint pens that observers in Frankfurt, Milan, New York, and Tokyo have recognized as a successful, individual and beautiful example of form in the worldwide history of writing instrument design.

The finished product: Easy to hold, with a three-sided front section.

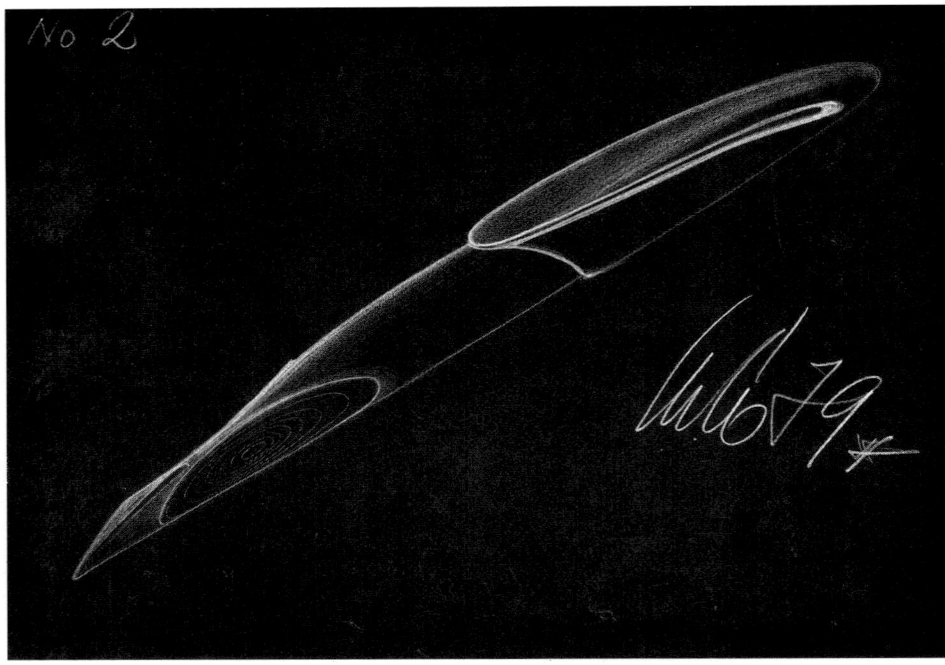

The first designs were made on black cardboard.

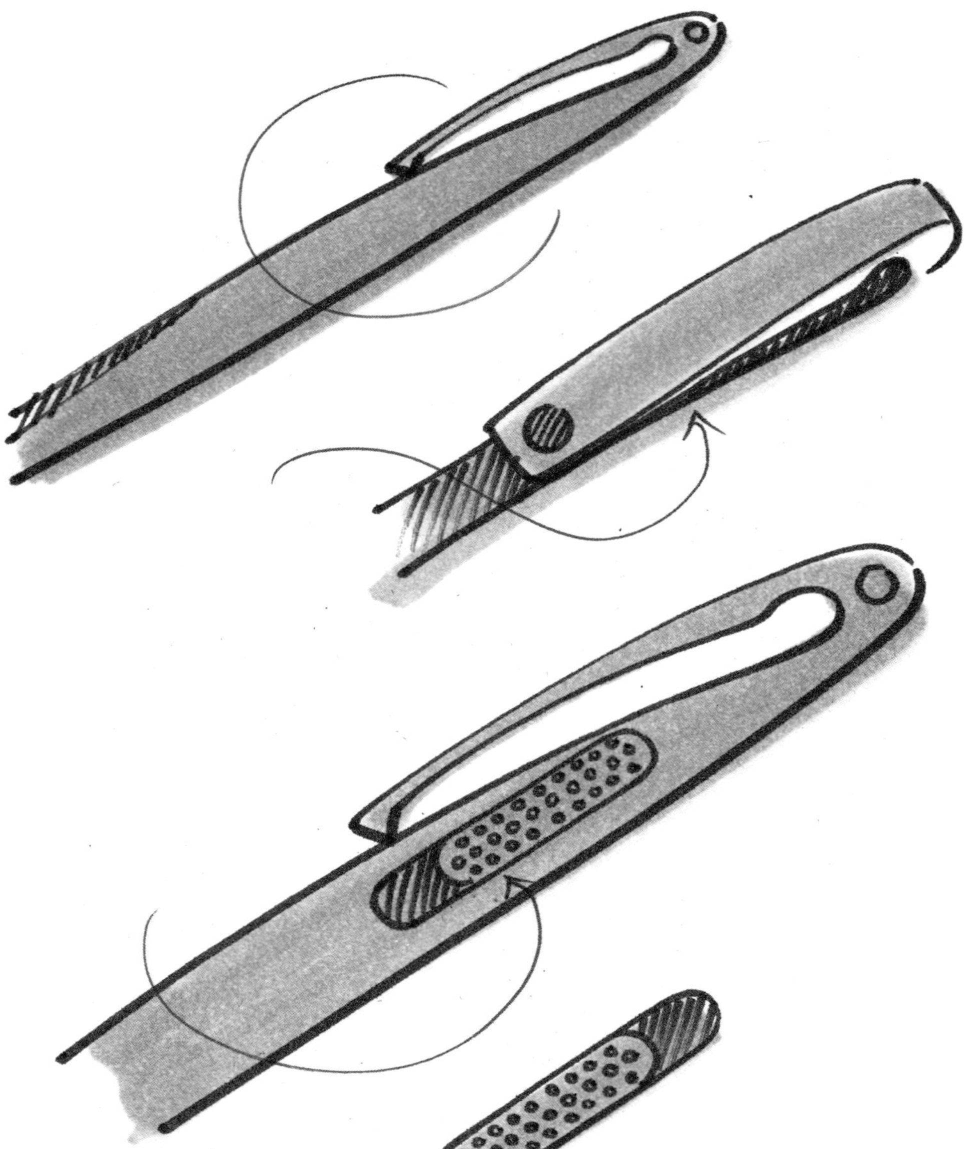

Detailed studies in a more advanced phase.

Senator

Fine mechanical work has been at home in the Odenwald for years. In 1920 the merchant Georg Merz, along with the respected master turner Justus Krell, established a small factory for writing instruments in Gross-Biberau. At first wooden penholders and the steel nibs that went with them were produced. In the Fischbach Valley there was an old established woodcarving industry from which the new firm acquired its personnel.

Merz & Krell also made a name for themselves in producing high-quality turned objects for industries that manufactured surgical instruments or decorative items. The former "Schuchmann Mill" in which the business had been located quickly grew into Germany's most significant producer of mechanical pencils and a noteworthy maker of piston fountain pens. Mechanical pencils in umbrella form, those that contained sewing equipment, or had a built-in perfume bottle as a "bottle in a pocket" became specialties of the house and gained great public popularity at the Frankfurt and Leipzig fairs.

In the 1930s, production of piston, safety, lever, and pressure filling fountain pens began. The favorite material was celluloid, out of solid blocks of which the individual parts of the pens were turned.

At the end of the 1950s the emphasis in production changed. The ballpoint pen had won millions of buyers. Merz & Krell looked back to their previous successes with their original mechanical pencils as advertising and promotional articles. Now similar ideas were carried over to modern manufacturing methods. Ballpoint pens were lettered in stamped foil, offset printing and finally screen printing. Having produced more than 75 million plastic writing instruments, Merz & Krell, with its Senator brand, holds a leading position among German writing instrument manufacturers today.

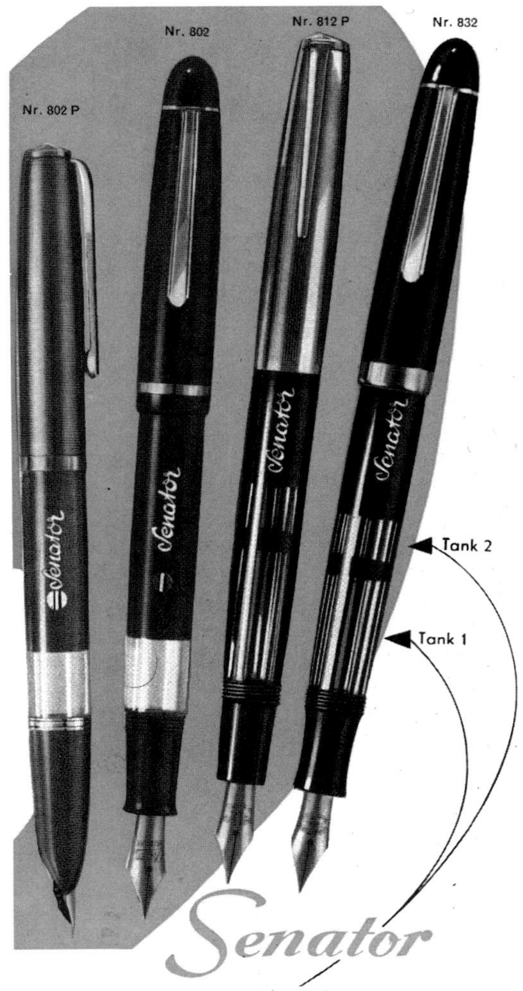

Senator fountain pens with doubled ink reservoirs, from the 1950s.

Fountain pens and mechanical pencils made of celluloid were sold under the "Melbi" brand name in 1932.

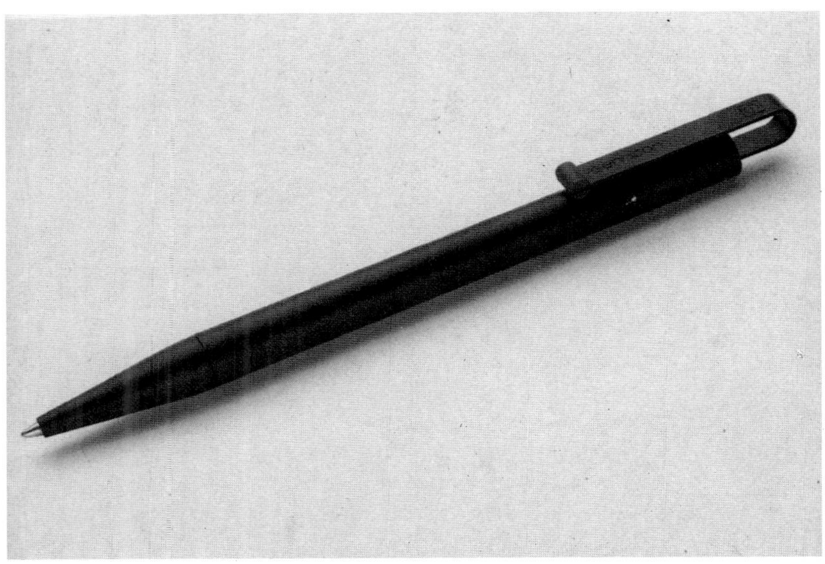

A modern "Avantgarde" Senator ballpoint pen with clip/push-button mechanism.

Almost a Museum: 6500 Ballpoint Pens Under one Roof

That is how many promotional ballpoint pens Georg Schriever-Abeln has collected under one roof—that of his Bremen home. The chief decorator of a department store branch got the idea of going into this hobby when he cleaned off his little daughter's desk. A good two dozen pens in striking shapes and colors fell into his hands and formed the foundation for an extraordinary collection. This devoted collector has obtained most of his most colorful specimens free. No taxi driver, no visitor is safe from the stubbornness of this collector, who comes from Oldenburg. Interesting pens are exchanged on the spot for duplicates from his own stock and find their way, carefully sorted, into thirty drawers or transparent plexiglas cases. Along with old-timers by Montblanc or Lamy in sober black, decorated with shimmering gold rings and clips, are the striking pieces that make the collector's heart beat faster. Lipstick tubes, potatoes, eggs, toothbrushes and toothpaste tubes, screws, rulers, tennis rackets, even a waffle iron had to be kept as original ballpoint pen shapes.

Souvenir shops in particular are sources of treasures for the dedicated ballpoint pen collector. Here products of the pen industry from Hong Kong, Taiwan, Korea, and Italy are found, waiting for the postcard writing public to buy them and take them home. The arrays of promotional items used by the international representatives of large industrial firms are just as rewarding. For them a promotional pen is a "door opener" to many a sales office. It is said that the more attention the promotional pen gets, the more successful the sales talk will be. As already noted, most of the models in this unique collection were obtained free. Only once has Georg Schriever-Abeln had to dig deep into his pocket: His most expensive pen, a ten-color ballpoint, cost $3.95 in New York. After his return he discovered the same model in a department store at home for 2.95 Marks . . .

About 6500 ballpoint pens in all imaginable colors and shapes have been collected by Georg Schriever-Abeln over the years.

A plate of colorful kulis.

Souvenir ballpoint pens . . .

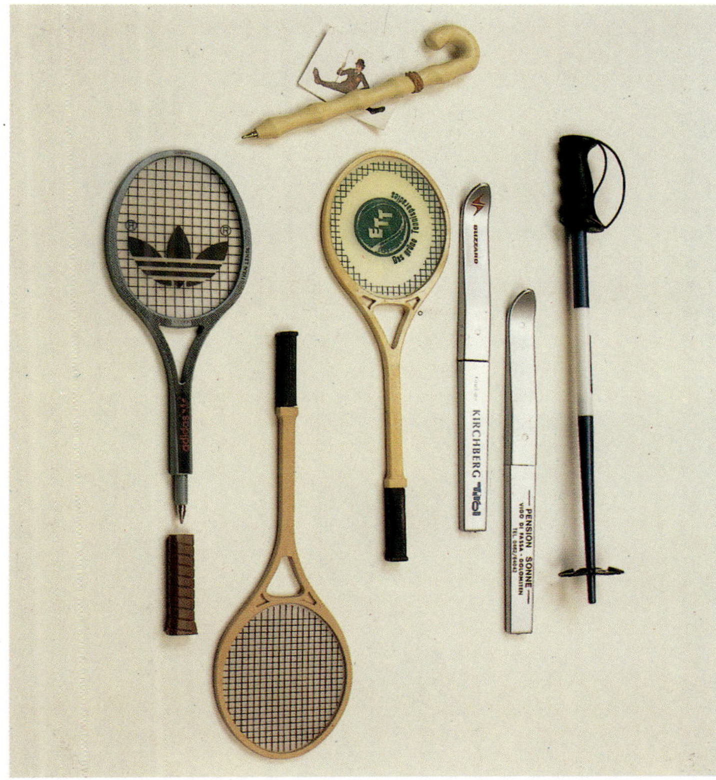

. . . advertising pens . . .

. . . lipsticks, toothbrushes and toothpaste.

Everything is present, even hand tools that write.

Felt and Fiber Tip Markers

One of the first European felt-tips, the Kaweco "Signier" pen of 1911.

"Mannenpitsu" is the Japanese word for "fountain pen." Freely translated it means "thousand-year paintbrush." Since antiquity, Chinese and Japanese writing has been done with a paintbrush. And the millennia-old wish of the Oriental peoples for a brush that would write "for years" was fulfilled in an incomplete way by the fountain pen and, later, the ballpoint pen brought by the American occupation forces. Their relatively fine lines may be suitable for the everyday Japanese for note-taking and everyday correspondence. But he can create expressive calligraphy only with the brush.

The invention of the felt-tip marker is usually attributed to the Japanese as having provided them with a "filled brush." The Kaweco felt-tip pen that was used in offices and workshops before World War I, at least in Germany and France, is not so well known. The honor of rediscovering the felt-tip for the Europeans belongs, beyond a doubt, to Volker D. Ledermann, who introduced the Japanese felt-tip in 1959 as "edding No. 1." It is still available today, and it works in a surprisingly simple way. In a barrel that comes to a point at the front there is an ink-saturated reservoir of wadding. As a result of capillary attraction, the ink is led from the reservoir to a woolen-felt writing tip. A simple but effective system. The excellent covering, quick-drying, watertight ink allows large-scale writing, drawing, sketching, and marking. The quickly evaporating liquid solution in the ink requires careful closing (capping) to avoid premature drying.

The large-size felt-tip marker for "coarse" writing and marking was followed in 1963 by the slim Pentel *Sign-Pen*, a fiber marker for writing and sketching. Its writing tip consisted only of strands of 4000 to 8000 synthetic fibers, running parallel to each other, that had been hardened. Bundles of nylon fibers produce very

Individual parts of a felt-tip marker: Closing cap, mouthpiece, barrel, closing plug, liner, ink reservoir of felt or fiber, writing tip.

long-lasting but coarse writing tips. Acrylic and polyester provide soft, easy writing but also wear comparatively fast. The fiber bundle is cut down to form a point. To retard damaging drying, the writing fluid contains a portion of glycol. This alcohol does two jobs. It absorbs water from the air and delays drying.

The fiber marker soon became a fashionable writing instrument. Although it does a bit of squeaking, one can write with it very well without problems—and of course one's ink writing looks more elegant than ordinary ballpoint writing. The disadvantage of not being able to press hard when, for example, filling out a Lotto form, was accepted calmly. And for children, the fiber-tip markers in 6, 12, 18, or even 36 colors gave a great deal of drawing pleasure, especially as in most cases the ink could be washed out of their clothing or gone over with an eraser pen in case of mistakes, usually with no difficulty.

The trend of the 1970s inclined toward fine writing. Stroke widths of 0.5 mm, 0.3 mm, or even 0.1 mm could be achieved, either by cutting down the resin-hardened bundle of fibers or by making the bundle itself smaller. Fiber markers do, however, have the natural tendency of becoming gradually wider and wider during the course of their writing lives.

The cross-section of an extruded plastic tip, as used for fine-writing fineliners. The ink is fed through the symmetrically arranged hair-fine capillaries.

Again it was the Japanese who made progress in felt- and fiber-tip development. They developed a plastic point, made by the extrusion process, that includes dozens of very fine capillary channels. The point keeps its form, even after long writing, as long as it is held and supported by a metal housing. The fineliner, so called because of its extremely fine stroke widths, first conquered the graphic art, design, and architectural offices in the 1970s, found its way into technical drawing offices and, in the 1980s, into the private home as well.

The areas in which felt-tip, fiber-tip and fineliner pens can be used are almost limitless. There are types that write on practically all surfaces, including glass, porcelain, metal, wood, and plastic. The graphic artist uses them for a fast scribble or layout, the engineer for construction sketches, the bookkeeper to enter numbers in narrow columns, the school pupil for drawing and learning to write, and the decorator for working out designs. Even the surgeon can use a special felt-tip for marking out operation procedures on the patient's skin. The multitude of uses have even given birth to special markers: the OHP marker for overhead projection, text-marking pens for stressing and locating important passages in books, and household types that do everything, even writing on aluminum foil wrappings of frozen food.

For the hobbyist the painting pen is a welcome helper. It contains well-covering pigmented inks that dry like paint, watertight, in the brightest colors. Unlike inks that contain dissolved coloring matter, pigment inks would clog the capillary passages of felt, fiber and plastic points. Painting pens thus have valve control instead of a capillary system. Pressure on the writing tip opens a valve and lets ink flow out of the tank (these models have no ink reservoir) and to the writing tip, which is thus saturated from inside to outside.

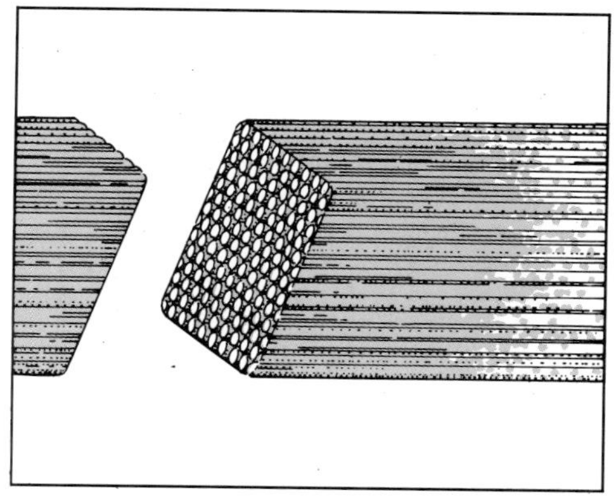

Fiber tips of polyester or nylon consist, depending on their diameter, of 4000 to 8000 fine synthetic fibers bundled parallel to each other. The cross-section shows how the ink flows between the fibers.

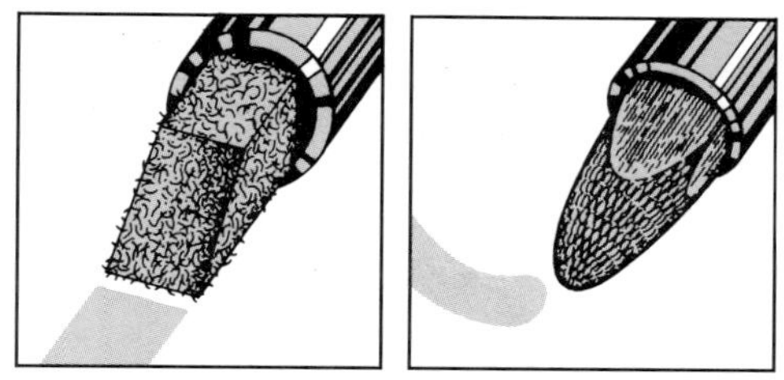

Felt writing tips are pressed of felt material and then stamped to the right size. The robust material gives a soft feeling when writing, but an even or sharp-edged script is not made by the felt-tip.

1

2
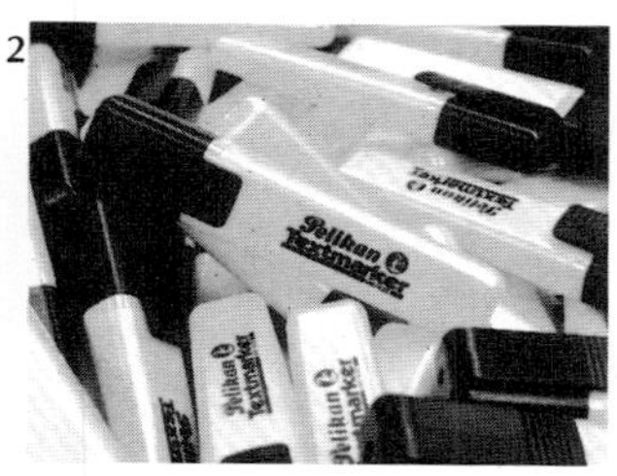

3

1. Fineliners being tested. With the help of a "writing automat" their evenness and length of writing are tested.

2. Markers for many types of use: Text markers . . .
3. . . . Overhead markers

Many felt-tip writers have changeable writing tips.

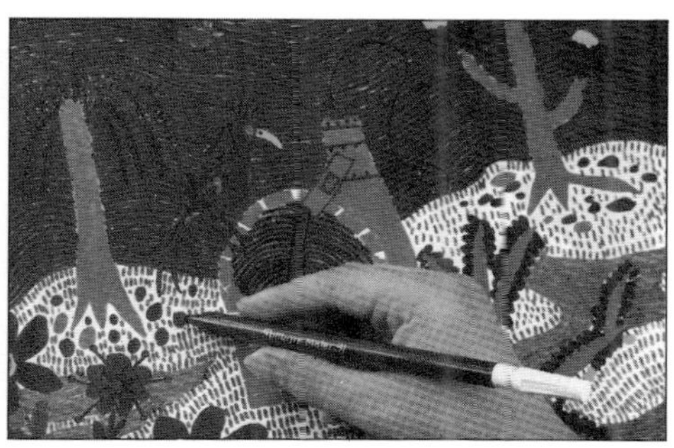

A pleasure for children: painting with the glowing colors of felt-tips.

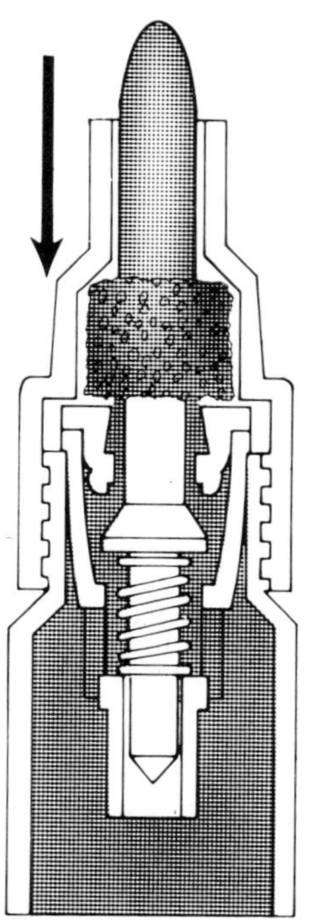

Felt-tip pens with valves are used for, among other things, paint touch-up sticks.

edding

On one of his numerous business trips in 1959, the Hamburg businessman Volker D. Ledermann noticed that his Japanese business contact was writing notes with an instrument hitherto completely unknown to him. The pen had neither a ball as a writing tip, like a ballpoint pen, nor a nib. He could see at a glance that it was not a grease pencil either, despite the vivid black writing. Curious, Ledermann asked to try out the unknown instrument. A point of felt put ink on the paper lightly, flowingly—like a brush filled with paint.

Volker D. Ledermann instantly recognized the many possibilities that this "ready paintbrush" offered. It was an ideal instrument for marking, writing, drawing, and sketching. The absence of any mechanism that could break down or wear out was another advantage. So he turned to his friend Carl-Wilhelm Edding, who was active in business with Japan at that time, and asked for his help. For one thing was clear to him: Such a writing instrument would change writing habits in many areas.

The *Lion Schnellschreiber* was the first felt-tip marker that the two imported into Germany. The new instrument was quickly adopted by graphic artists, office workers, and shipping departments of industries. To maintain success over a long period, they created a brand name. The *edding No. 1* midget felt-tip formed the foundation for the new company's quick expansion.

Millions of felt- and fiber-tip markers are sold all over the world today. The best-known of them may well be the *edding 3000*. This product, presently in production, is of almost unlimited use by the layman. Pens with corrosion-free indelible ink help the aircraft manufacturer or are used for special tasks at atomic power stations. Fiber-tip pens with document-true indelible inks preserve what would otherwise fall prey to time, while felt-tips add invisible identifying marks to valuable objects to protect against theft, providing a new tool for criminologists. Markers with weatherfast indelible ink, touch-up markers with indelible inks in wood-finish colors to cover scratches on furniture, and many other models fulfill even the most unlikely functions.

The legendary edding No. 1 is still in production today.

Different writing systems: from the solid metal point for the finest lines, only 0.5 mm wide . . .

. . . to the fiber tip for writing and sketching . . .

. . . to the felt-tip with angled point, that makes broad, powerful lines.

How an Ideal Pen should be Designed

Mel Strohminger of Baltimore owns one of the largest collections of writing instruments in the USA. This management consultant and father of four children writes commentaries, essays, and interviews for international journals.

"I have been collecting pens for about twenty years, simple ones with a goose quill as a writing tip, for dipping into an inkwell, insertion pens, safety fillers, piston fillers and cartridge holders. And I have gotten into the habit of writing with all of them, at least for a while, at the office and at home. It interests me to see what advantages and weaknesses can be found in each one.

"If you ask me which pens from which era look best to me, I'll tell you that about six fountain pens are very good. There is not yet an ideal fountain pen. I'll try to describe it.

"First of all, I'm speaking only of fountain pens, because I consider them superior to all other writing instruments in terms of handwriting. No ballpoint pen, no felt-tip can reproduce handwriting in its thick and thin lines (as the American "Copperplate" or the German "Sütterlin" demand) as well as a steel or gold nib, and it doesn't matter if the nib is made of 14- or 18-karat gold.

"In simple terms there is a relationship between the flexibility of the nib and the total weight. In general, a flexible nib allows a lighter pen, because the writer can then use the pen with a lot of feeling. The harder the nib—that means the less easily the halves of the nib spread under pressure—the heavier the pen can be. As a general rule: the more flexible the nib, the lighter the pen. And I'd like to qualify that by saying that even with a very stiff nib, it is tiring to write, even with a heavy pen.

"How long should the ideal pen be? Most Americans carry their fountain pen in their shirt pocket; the Europeans prefer the inside jacket pocket. If one wants to carry his pen in his shirt pocket often, then it is important that it is not too long or too heavy.

"Many fountain pen fans speak enthusiastically of the ideal position of the pen's center of gravity. I prefer to ask: How does the pen feel in your hand? You get a feeling for the pen with its cap on the end when you write with it. Either you feel that it suits you and fits in your hand, or not. It doesn't really matter if it is carefully balanced in the middle. A fountain pen designer who developed a very famous pen for a leading manufacturer told me this. For a test, a sample was chosen out of a row of prototypes of varying size, shape, and weight. This model was given to test people who sat in a dark room and could feel the pen and write with it. In this way they had an easier time of deciding which of the pens "felt best." I preferred a sharply slit nib that even cut into the paper a bit, in order to make clear, strong lines. In addition, the resistance decreased the speed so much that this helped make the writing clear. Nibs that are cut too round give a too-soft, "gliding" feeling when writing. They are just as hard to control as, for example, a ballpoint pen.

"What must be done to find an ideal fountain pen? My answer is to take your time in making a choice. Visit several shops and write with as many models as possible, so you can develop a basis for comparison. And so you can keep one of the variables constant, take along the paper that you use for your daily writing.

"My testing method is this: Write three or four pages quickly and with normal pressure. Does the pen skip? Does your hand get tired? Then the model certainly should not be considered. I am of the opinion that the grip section has to be long enough to give enough room for the index finger, thumb and middle finger. It should be round and have a recessed area, rather like a soft drink bottle, to allow different holding positions and not let the fingers slide off.

"In my opinion the cap with a screw thread is superior to the slip-on cap that is simply pulled off. The screw cap also offers more protection against unintentional loosening, for example, in the shirt or jacket pocket.

"I would prefer a clip with adjustable tension. That is not the case with most modern pens. Bend it back about half a centimeter. It should spring back to its original position without losing tension.

"And finally, a word about the filling system. Today the piston and cartridge filling systems are most widespread. Both have their advantages: The piston filler holds more ink, the cartridge filler is easier to fill, which is especially convenient when traveling, since one doesn't have to take an ink bottle along. I prefer traditional filling with a precise filling mechanism. It simply goes better with a valuable fountain pen—you wouldn't buy a luxury car with a small-car motor.

"As for the beauty of a fountain pen, I would give you this advice: Choose a timeless, elegant, not overdone model. A good pen can stay with you a lifetime and in time will become a useful utensil that expresses a part of your personality."

Mel Strohminger

Historic and Modern

The number of fountain pen brands and models made during the last eighty years is all but incalculable. There is much to collect, and many a collector's item has lain untouched in a desk drawer for many years until the original owner's descendants found it and made it available, while many more have been preserved and offered for sale by junk-shop owners or antique dealers. In addition to collecting them, many collectors enjoy writing with a venerable or even historic fountain pen.

1931

Early German Brands (no longer in production)

Argument-Füllhalterfabrik, Georg Karasch, Berlin
Astoria-Füllhalterfabrik, Hamburg
Greico, Otto Greiner & Co., Leipzig
Greif-Werke, Goslar
Goldfink (Trade name), Berlin
Haro-Füllhalterfabrik, Hans Roggenbuck & Co., Frankenstein, Silesia
A. W. Faber-Castell, Stein, near Nürnberg
Heiko, Walter Heise KG, Halle (today VEB)
Heintze & Blankertz, Berlin
Imperial, Gerlach & Bezner, Leipzig
Kaweco, Badische Füllhalterfabrik, Wiesloch
Luxor Pen GmbH, Baden-Baden
Marathon-Werk, Ernst Dietrich & Co., Schwäbisch Gmünd
Matador Füllhalterfabrik, Siebert & Löwen, Wuppertal-Elberfeld
Mars Bleistift-Füllhalterfabrik, J. S. Staedtler, Nürnberg.
Neco, Nehrlich & Co., Lingen/Ems
Nio, Robert Niehaus, Osnabrück
Osmia Füllhalterfabrik GmbH, Dossenheim
Rakete Füllhalterfabrik, Meissner Brothers, Beerfelden. Hessen
Röder Stahlzackwerk, Berlin
F. Soennecken Schreibwarenfabrik, Bonn
Uhu, Lingner & Fischer, Baden-Baden
Wegu Füllhalterfabrik, Willy Gunkel, Leipzig
Welt-Pen GmbH (Favorit and Rex brands), Berlin

Other Brands

Germany

1931

Hermann Böhler GmbH, Dossenheim
Brause & Co., Iserlohn
Elysée, R. Dummert GmbH & Co KG, Pforzheim
Geha-Werke, Hannover
Goldring Ludwig Schleif, Bühl, Baden
Hauser & Co., Gosheim
IMCO, J. Michaelis (Diplomat) GmbH, Stuttgart
Klio-Eterna GmbH, Wolfach
Kreuzer, Hannover
Lamy, Heidelberg
Laurin/Rebhan, Hannover
Merz & Krell (Melbi, Senator), Gross-Biberau
Montblanc-Simplo, Hamburg
Pelikan, Hannover
Porsche Design (Faber-Castell)

Pen Brands

Reform, Mutschler Schreib- und Zeichengeräte GmbH, Heidelberg
J. Schröder (Tropen), Lüdenscheid

1949

Belgium

Le Tipa
Mercury

France

Bayard
Conté
Pierre Cardin
Cartier
Christian Dior
Daniel Hechter
Yves Saint-Laurent
J.M. Paillard
Reynolds
Waterman

Italy

Aurora
Ancora
Berger & Wirth
Edacoto
Ferrari
Gold Star
Montegrappa
Omas
Stilu
Tabo
Tibaldi
Victoria
Yaffa
YES OK

Great Britain

Army & Navy Cooperative Society Ltd.
Conway Stewart
Dunhill
Esterbrook
Mabie, Todd & Co. (Swan)
Mentmore (Platignum)
Thomas de la Rue (Onoto, Pelican)
Perry & Co. (Osmiroid)
W.H. Smith & Son Ltd.
Unique Fountain Pen
John Whytwarth
Wyvern Fountain Pen

Switzerland

Caran d'Ache

Spain

Inoxchrom

Japan

The countries of the Far East, China, Korea and Japan, have a tradition of classic handwriting. To this day it has not been possible to produce the complex Japanese characters of the three written systems, "Chinese script," "Hiragana," and "Katagana," satisfactorily with a typewriter. Thus it is natural for the highly developed Japanese economy and government to carry on their daily correspondence by handwriting. It is no wonder that Japan leads the world in per capita consumption of writing instruments. Especially prized as status symbols are high quality fountain pens and ballpoint pens from Germany, France and the USA.

At the beginning of recorded history, the writing brush came to Japan from China. In the Middle Ages the Samurai, the knights of old Japan, used a small inkstone along with the writing brush and carried them in a quiver, the Yatate. These portable writing tools of iron, brass, copper, bronze, silver, horn, or bamboo were decorated with inlay work and often colorfully lacquered. With the help of the Yatate, the Samurai wrote reports to their commanders from the battlefield or letters to their families. Only toward the end of the 19th century did the first fountain pens from the West reach Japan: the American Caw, British Swan and Onoto.

1922

Pilot

In 1918 the seafaring firm of Namiki and Wade founded a fountain pen factory, which they named Pilot. To protect their hard rubber pens from drying out, they covered them with silver or gold lacquer in the tradition of Japanese handicrafts. The expensive writing instruments found so much acceptance, especially in England, that in 1930 Dunhill decided to offer the "Dunhill-Namiki" exclusively in Europe. It is still regarded as one of the world's mostg instruments.

Platinum

Since costly platinum metal was an important component of his pens' nibs, S. Nakada named his new company "Platinum" in 1919. Impressed by the quality of the Waterman pens, S. Nakada at firstbegan to build similar ones under the *Global Mark* name. Several successful designs subsequently made Platinum one of Japan's largest manufacturers of writing instruments.

Sailor

At the age of fourteen, K. Sakata decided to build a fountain pen of the type that a friend has brought back from a trip to England. In 1940 Sailor produced the first Japanese ballpoint pen. In 1971 the firm began the exclusive sale of Sheaffer fountain pens in Japan. The *Trident,* introduced in 1983, is regarded as an innovative fountain pen. Instead of a nib, it has three sharp-pointed triangles of spring steel welded together. Like a ballpoint pen, the *Trident* allows writing in all directions.

USA

Aiken Lambert
Auto-Filler
Camel
Carter
Caw
Chilton
Conklin
A. T. Cross
Dunn
Eagle
Eclipse
Edison
Foley
Frankling
Gold Medal
Grieshaber
John Holland
Independent
Inkograph
Juco
Keystone
Laughlin
LeBoef
Lincoln
MacKinnon
Marvel
Meredith
Moore
Parker
Post
Sheaffer
Venus
Wahl-Eversharp
A.A. Waterman
L.E. Waterman
Paul E. Wirth
Weidlich
Windsor

1903

Index

ABS, 129
Advertising pens, 166
Album, 13
Alizarin, 20
Alphabet, 10
Altamira, 6
Arrow, 79
Art-Deco, 73, 88, 91, 148, 150
Aurora, 140, 149
Automatic, 31, 76, 144
Automatic-Pen, 141
Artos, 107

Bakelite, 128
Balance-look, 91
Balance Pen, 86, 149
Ballpoint cartridge, 160
Ballpoint pen, 93, 106 ff, 160, 165
Band pens, 17, 44
Bayard, 42, 92
Big Red, 78
Billet-Stift, 49
Birmingham, 36
Bismarck pen, 37
Biró, 79, 94, 106, 147, 159
Blackbird, 147
Blanckertz, 36
Blanzy, 32
Blimpy, 79
Brause, 124, 144
Brown, 77

Calamus, 11
Calligraphy, 14, 17
Camel, 58, 126, 140
Cap, 150
Capillary action, 68, 70, 72, 75, 138, 157
Capillary channels, 68 ff
Capillary principle, 68
Capillary system, 145
Capitalis, 11
Carrera, 101
Carter, 84
Caw, 74, 140
Cedar, 49, 52
Cellomine, 62
Celluloid, 86, 128, 129, 154
Cellulose, 88
CF, 73
China lacquer, 131
Clip, 150
Club des Collecionneurs de Plumes, 125
Coin filler, 141
Colani, 163 ff
Colored pencil, 46
Conklin, 76, 84, 118, 126, 141 ff, 150
Connoisseur, 87
Contè, 49, 90
Converter, 145
Conway-Stewart, 84, 89, 93, 118, 152 ff
Copying pencil, 62
Cosmetic pencil, 62
Crescent filler, 76, 86
Cross, 49, 102, 149, 162
Cumberland, 46
Cuneiform, 7

Design, 163
Devotional pens, 37
Dia, 97
Diamond slit, 130
Diamond point, 37

Dinkie, 152
Diplomat, 107, 145, 149
Document-true, 160
Drawing ink, 8, 9, 103, 104
Drawing pen, 105
Duofold, 78 ff, 88
DuPont, 88, 149
Doric, 91, 94, 149
Duro, 152
Dunhill, 149

Ebonite, 88, 127, 138
Edison, Thomas A., 74
Electroplating 130
Elite, 97
Endura, 150
Eterpen, 94, 105, 156 ff
Even-stroke pen, 17
Eversharp, 49, 79, 94, 136, 140, 144, 149
Eyedropper, 69, 76 ff, 139 ff
Eyedropper filling, 140, 146

Faber, 49 ff, 54, 94, 157, 159
Faber, A. W., 54
Faber, Eberhard, 54, 60, 106, 155
Faber-Castell, 54, 107, 124, 140, 144
Faber, Kaspar, 54
Fagard, Jules, 70, 72
Fall pencil, 50
Feather Touch, 86
Felt-tip pen, 96, 168, 170
Fiber pen, 168
Fifth Avenue, 94
Filler pencil, 50
Fine lead pencil, 50
Fine-liner, 169
Fisher Price pen, 159
Fixed pipe, 50
Flint tool, 6
Forms, 148
Fölsch, F. B., 67
Fountain pen, 172
Fox, S., 68
Full Moon pencils, 60

Gabirelli, 149
Galalite, 128 ff
Gall, 18
Geha, 138 ff, 144 ff
Gentleman, 73
Gillot Joseph, 32
Glass pen, 101
Gold, 129
Gold plating, 130
Goldfaber, 56
Goldflash, 130
Goldschwinge, 124
Gomez, Francine X, 73
Goose feather, 32
Graduate, 73
Graphite, 46, 52
Graphology, 41
Greif, 107, 124
Guilloche, 130
Gutenberg, 14
Guttknecht, 49

Half Slide, 50
Half-unzial, 12
Hard rubber, 69, 127 ff
Hardness, grades of, 49
Hardtmuth, 49

Hieroglyphics, 7
High-polymer, 50, 60
His Nibs, 124
Holland, John, 118
Hundred Year Pen, 70

Ibis, 23
Ideal Regular Fountain Pen, 69, 74
Ink, 18
Ink cartridge, 72 ff, 87, 145
Ink equalization, 70
Ink feed, 138
Ink filler, 102 ff
Ink Kuli, 102, 104 ff
Ink molecule, 72
Ink passage, 68 ff, 75 ff, 78, 138
Ink pencil, 102
Ink roller, 159 ff
Ink tablet, 84
Ink Vue, 144
Ink writer, 102
Insertion filler, 72, 77, 139
International Forum, 128
Inventions, 82
Iridium, 135
Iron-gall ink, 18

Jackdaw, 147
Jack-Knife, 128
Jif, 22, 73
Jointed pen, 50
Jotter, 79, 106, 158 ff

Kaweco, 91, 93, 102, 107, 118, 124, 140, 143 ff
Keystone, 74
Klingspor Museum, 17
Kolleg, 96 ff
Kurz, 49

Lady Patricia, 73
Lamy, 124, 145, 149, 151
Lascaux, 6
Lászlo Biró, 156
Laughlin, 74
Lead-white, 46, 48
Leonardi, August, 18, 20
Lever, 84
Lever mechanism, 84
Lever pen, 42, 84, 142
Lifetime, 86, 148
Lincoln, 74
Logwood ink, 20
Lucens Extra, 90
Lucite, 129
Lucky Curve, 75, 78, 138
Luna, 58
Luxor pen, 79
Ly-pen, 37
Lyra, 49

Mabie, Todd & Co., 89, 146
Maginnis, James P., 69 ff
Maginnis, P, 64
Magna, 133
Mallat, 37, 90
Man 100, 73
Mars, 58
Matador, 107
Mechanical pencil, 49, 162
Meisterstück, 100 ff, 107
Melbi, 90
Membrane filling system, 144

Meredith, 74
Merz & Krell, 107, 144
Mitchell, John, 32
Minor, 133
Minuscule, 11 ff
Monte-Rosa, 101
Montblanc, 89, 102, 103, 107, 118, 123, 126, 140, 142 ff, 145, 149 ff
Montblanc Simplo, 98
Moon Pencil Factory, 58
Moore, 118
Moseley, Walter, 65, 67

Newel A. Prince, 67
Nib widths, 136
Noris, 58
Normal pen, 36
Nostalgia, 87
Nozac, 144

OMAS, 90, 126, 150
Onoto, 74, 76, 84, 89, 90, 106, 118, 126, 132 ff
Osmia, 56, 89, 107, 118, 124, 144
Osmium, 136
Othello, 62

Paillard, 90
Pantabille, 73
Paper, 10, 18
Papermaking, 9
Papermate, 158, 160
Papyrus, 8, 10 ff, 18, 46
Parchment, 10, 12, 18, 46
Parker, 25, 51, 65, 75, 79, 81, 84, 88, 89, 91 ff, 106, 118, 126, 133, 137 ff, 142 ff, 145, 148 ff, 150 ff
Parker, George S., 74, 78
Parker, John J., 67
Patent, 82
Patrician, 73, 148
Pelican, 22, 89 ff
Pelikan, 22, 84, 89, 93, 107, 118, 126, 132, 144 ff
Pelikan Ibis, 122
Pelikano, 23
Pen Fanciers Club, 124
Pen World, 124
Pencil, 46, 49, 52
Perkeo, 96
Permanite, 129
Perry, James, 32
Picture postcards, 39
Pilot, 84, 142, 174
Pilot-Pen, 90
Piston filling system, 23
Piston pen, 67, 143 ff
Piston pump, 77
Plexiglas, 129
Platinum, 84, 90, 142, 174
Poetry, 150
Polystyrol, 129
Portrait pen, 37
Post-pen, 76, 143
Präsident, 42, 89
Premier, 79
Pressure pen, 143
Pressure pencil, 49, 50
Prince, 65
Progress, 56
Pulton, George, 67
Push filling, 143
PVC, 129
Pyralene, 127, 129
Pyroxylene, 86, 88, 94, 129

Radite, 88, 129
Rappen, 23, 89, 140
Redis-Feder, 37
Reflex, 101
Reform, 145
Regent, 124
Regular Pens, 84
Regulator, 138
Reynolds, 106, 157
Rheingold, 42, 89
Riepe, 103 ff
Ripple, 148
Roeder, 142
Rolled gold, 130
Ronson, 126
Rosetta Stone, 7
rotring, 104 ff
rotring-Rapidograph, 103 ff
Rouge et Noir, 99
Runes, 11

Safety pen, 42, 72, 77, 81, 84, 140
Sailor, 84, 142, 174
Salvation Army pen, 76
Satin-Flo, 106, 157
Schneider cartridge, 159
Schwan, 49
Schwanhäusser, Gustav, 62
Schwan-Stabilo, 62
Schwentner, Daniel, 67
Screw cap, 150
Seal, 24
Self-filler, 76
Self-filling lever type, 84
Senator, 145, 165
Senefelder, Aloys, 32
Shaw, Robert, 68
Sheaffer, 84, 88, 91, 106, 118, 126, 138, 142 ff, 149 ff
Sheaffer, W. A., 86
Siegert, H., 65, 67
Signier fountain pen, 96
Silver, 129 ff
Silver stylus, 146
Silver-Sleeve, 130
Simplo, 89
Simplo-Filler Pen Co., 98 ff
Skyline, 92, 144
Slim-Line, 149
Slip-on cap, 150
Snorkelpen, 87, 106, 138
Soennecken, 36, 39, 42, 44, 62, 74, 89, 93, 107, 118, 139 ff, 142 ff, 148, 151
Souvenir pens, 37
Souverän, 24
Space Pen, 159
Spear Head, 138
Spieker, 18
Stabilo, 62
Stabilo Boss, 62
Staedtler, 49, 107, 144
Staedtler, J. S., 58
Stainless steel, 130
Steel pens, 32
Sterling silver, 130
Stewart, 118
Stratowriter, 87, 106
Stylo, 11, 96, 102
Stylograph, 102
Stylopen, 104
Stylos, 100
Stylus, 11

Surface tension, 68
Swallow, 96
Swan, 74, 84, 89 ff, 106, 118, 140, 142, 146 ff
Swan Pencil Co., 62
Swano, 62
Swift, 132
Swinger, 79

Tablet pen, 140
Tally stick, 12
Tannin, 18
Targa, 87
Tempoint, 94
Thomas de la Rue, 22, 76, 89, 90, 132, 142
TK-matic, 56
To-Feder, 37
Toledo, 23 ff
Touch-Down, 87, 106, 143
Trident, 174
Triumph, 86
Tube filler, 141
Tuckaway, 86
Twist filling, 142

Unzial script, 12
Urushiol, 131

Vacumatic, 78, 80, 91 ff, 144
Visofil, 147
Vulcanite, 127

Wagner, Günther, 22, 89
Wahl-Eversharp, 84, 89, 91 ff, 106, 118, 142, 150, 156 ff
Wasp, 86
Waterman, 72, 76, 84, 88, 89, 91, 118, 126, 128, 138, 140, 142 ff, 144, 148, 150
Waterman's Ideal Fountain Pen, 72
Waterman-Jif, 72 ff, 90, 145
Waterman, L. E., 68
Waterman's Regular Pen, 77
Whytwarth, 119, 126
Wirth, Paul E., 74
Witt, de, 74
Writer's cramp, 39
Writing ball, 134
Writing Equipment Society, 124
Wyvern, 140

Youth Style, 148